OPHRYS

THE BEE ORCHIDS OF EUROPE

OPHRYS

THE BEE ORCHIDS OF EUROPE

Henrik Ærenlund Pedersen & Niels Faurholdt

Including *Ophrys* cultivation
by Richard L. Manuel

Kew Publishing
Royal Botanic Gardens, Kew

PLANTS PEOPLE
POSSIBILITIES

First published in 2007 by
Royal Botanic Gardens, Kew
Richmond, Surrey, TW9 3AB, UK

www.kew.org

ISBN 978 1 84246 152 5

British Library Cataloguing in Publication Data
A catalogue record for this book is available from the British Library

COPY EDITOR: Alison Rix
DESIGN, TYPESETTING AND PAGE LAYOUT: Jill Bryan

FRONT COVER PHOTOGRAPH: *Ophrys insectifera* s.s. with its pollinator, *Argogorytes mystaceus*

PHOTOGRAPHY AND ILLUSTRATION CREDITS:
Bent Thøger Christensen 114; Niels Faurholdt half title, main title, 29, 33, 35, 36, 38, 39, 41,
48, 49, 72, 75, 80, 84, 85, 89, 94, 97, 99, 108, 109, 110, 113, 117, 123, 124, 129, 131, 133,
135, 141, 144, 145, 146, 150, 152, 154, 156, 173, 180, 181, 184, 187, 190, 193, 196, 197, 207,
212, 213, 220, 221, 223, 229, 250, 251, 252, 256, 260, 262, 263, 264, 270, 271; Cesario Giotta
132; Aase Gøthgen 72, 80, 124, 136, 159, 160, 161, 170, 188, 189, 190, 200, 205, 207, 217;
Linda Gurr 246, 247; Erik Jensen 165; CAJ Kreutz 76, 77, 142, 162, 163, 178, 179, 209, 213,
227; Jimmy Lassen cover, 3, 5, 12, 15, 16, 17, 19, 23, 25, 26, 47, 51, 57, end papers; Richard
L. Manuel 238, 239, 241, 244, 246, 247, 249; Bente Olsen 164; Henrik Ærenlund Pedersen 30,
32, 34, 37, 42, 43, 66, 67, 72, 73, 75, 79, 84, 85, 86, 87, 88, 89, 90, 91, 94, 95, 96, 98, 99,
103, 106, 107, 110, 112, 113, 116, 117, 122, 123, 128, 129, 130, 133, 136, 145, 148, 149, 151,
152, 154, 164, 169, 172, 173, 178, 183, 184, 185, 192, 194, 197, 201, 204, 206, 209, 212, 220,
221, 222, 223, 226, 230, 236, 253, 255, 258, 267, 268, back cover.

Printed in Italy by Printer Trento

For information or to purchase all Kew titles please visit
www.kewbooks.com or email publishing@kew.org

All proceeds go to support Kew's work in saving the world's plants for life

CONTENTS

ACKNOWLEDGEMENTS

We wish to express our special thanks to Jimmy Lassen for preparing most of the drawings and for trimming and adjusting image files. We are likewise grateful to Hans-Erik Bangslund for scanning the original colour slides, to Lene Kolind Skougaard for preparing the distribution maps based on our drafts, to Ingeborg Nielsen for trimming and adjusting image files, to Jean Schollert for technical support and to Lydia White and colleagues (Kew Publishing) for carefully seeing this book through to print.

We are indebted to Lars Vilhelmsen (Zoological Museum, Natural History Museum of Denmark) for placing preserved insects at Jimmy Lassen's disposal, and we are very grateful to Cesario Giotta, Aase Gøthgen, the late Erik Jensen, Karel Kreutz, Jimmy Lassen, Richard L. Manuel, Bente Olsen and Bent Thøger Christensen for allowing us to publish their excellent *Ophrys* photos. Svante Malmgren, Bent Vestergaard Petersen, Gerhard Stimpfl and Jan Frits Veldkamp have sent us useful information, and Carsten Koch has readily supplied us with foreign literature, whereas Phillip Cribb, Ib Friis and Kai Larsen have kindly promoted the manuscript. We owe all of them our grateful thanks.

Special thanks are due to Richard M. Bateman for carefully reading the manuscript and suggesting scientific and linguistic changes that greatly improved the text. We are also very grateful to Richard L. Manuel for contributing the chapter on *Ophrys* cultivation.

Last but not least, we wish to thank our many travel companions in the Mediterranean for fruitful discussions and stimulating company over many years, especially during the decade that has elapsed since we started work on this book.

INTRODUCTION

The bee orchids (the genus *Ophrys*) belong to the orchid family (Orchidaceae). This plant family is probably the largest in the world with at least 25,000 species divided among more than 850 genera. Only the family of composites (Asteraceae) might emulate the orchid family in species richness.

The orchids differ from most other plants because their variation and biological adaptations are so often strikingly eye-catching; within the orchid family itself, however, certain groups of species stand out among their relatives. In recent years, the charismatic bee orchids have caught the fancy of local residents, and have attracted many visitors to the Mediterranean each spring.

The pollination biology of bee orchids is fascinating. In many species, floral scents, structures and colouration patterns have evolved that are so reminiscent of the females of a bee, wasp or beetle species that they attract males of the same insect. The flower is pollinated when the sexually excited male insect attempts to copulate with it! The floral imitation of insects is species-specific to a certain extent, and because of this the pollination biology contributes to maintaining the species and subspecies of *Ophrys*.

However, it frequently happens that hybrids or isolated *Ophrys* populations start to run through a speciation process. At present, all imaginable intermediate stages exist between distinctly separate species at the one end of the spectrum to strongly variable, but still coherent, species at the other. This phenomenon makes *Ophrys* highly interesting for the study of evolution but, at the same time, the very vague separation of many entities has caused remarkable disagreements between systematists trying to classify the genus. Contrasting patterns between different kinds of available data have also contributed greatly to this confusion.

This book is a narrative of the fascinating natural history of the European bee orchids, based on an introduction to their structure and their occurrence in the wild. Additionally, it presents a classification based on an unusually wide species concept that is in sharp contrast to the narrow species concepts applied in currently prevailing classifications of *Ophrys*. We believe that our concept is more consistent with the biological background and that much of the morphological and biological variation in the genus is more appropriately recognised at the levels of subspecies and variety. We also think that our classification offers the beginner a more manageable survey of the species and their most essential variation patterns. We hope that the more experienced reader will be able to find inspiration by acquainting him- or herself with the ideas behind the classification (as we believe that at least some of the ideas can be utilised in connection with more profound research into the systematics of *Ophrys*). The book contains identification keys at both species and subspecies level, thus allowing readers to adjust use of the keys according to their own background and ambitions.

It is our sincere hope that this book will contribute to promoting interest in, and understanding of, the bee orchids of Europe – and in this way support the protection and management of these captivating plants in their natural environment.

REFERENCES: Cribb & Govaerts (2005), Pridgeon et al. (2001), Sundermann (1977).

1. Origin and systematic position

Being easily perishable herbs with tiny, quickly degradable seeds and firmly assembled pollen grains dispersed by insects, the orchids make very poor candidates for being preserved as recognisable fossils. This problem is further increased by the fact that most orchid species are found in the wet tropics (with quick decomposition) where they mainly grow epiphytically in trees, well separated from the aqueous environments that most effectively promote the fossilisation process. Indeed, not a single reliable orchid fossil is known, only a few dubious ones.

Today, orchids occur in all continents but Antarctica. However, the occurrences are very unevenly distributed. By far the largest number of species is found in the tropics, many fewer grow in temperate areas, while arctic areas accommodate very few orchid species. The genus *Ophrys* belongs to the *Orchis* subfamily (Orchidoideae), which is mainly distributed in temperate and subtropical areas in both the northern and the southern hemisphere. By far the major part of European orchid genera are members of this subfamily, and most of them, including *Ophrys*, belong to the *Orchis* subtribe (Orchidinae) of the *Orchis* tribe (Orchideae).

Phylogenetic relationships within the *Orchis* subtribe have been vigorously discussed for many years. However, DNA studies from the mid-1990s onward have greatly improved our understanding. As far as *Ophrys* is concerned, the results indicate that this genus is monophyletic (meaning that all bee orchids have one common ancestor and that all descendants of this ancestor are bee orchids). Additionally, the results demonstrate that the closest relatives of bee orchids are the genera *Serapias* and *Anacamptis* (in a wide sense, including several species previously assigned to *Orchis*). These findings are consistent with recently published cytogenetic data.

Despite the lack of orchid fossils, it is possible to develop a qualified hypothesis of the approximate age of the orchid family and its subfamilies. As changes in gene sequences over time occur at a relatively constant rate, DNA-based analyses can be calibrated with reliably dated fossils. Thus, in analyses of major systematic groups, the fossil record of some families can indirectly provide dates for families that are devoid of fossils. Recent application of this "molecular clock" approach has indicated that the orchids are one of the oldest families of monocotyledonous plants. The family apparently originated in the present Australasian region more than 100 million years ago; all currently recognised subfamilies were dated to be more than 65 million years old.

How and when the genus *Ophrys* branched off is unknown, but its origin (or early radiation) may have been closely associated with the evolution of pollination by pseudocopulation, which is an almost universal feature throughout the genus (see the section, From pollination to seed dispersal p. 24). The behaviourally active compounds in the scent of *Ophrys* flowers generally occur in at least trace amounts in plant cuticular waxes. As proposed by F. P. Schiestl and co-workers, the specific complex patterns of otherwise common compounds in the floral scent of *Ophrys* may have originated through occasional mutations that changed the relative amounts of the individual compounds. Representing successful adaptations for olfactory attraction and sexual deception of pollinators, accumulating mutations of this kind may have become fixed by natural selection, gradually leading to a highly specialised breeding system.

REFERENCES: Bateman et al. (2003), Bremer & Janssen (2006), Chase (2005), D'Emerico et al. (2005), Dressler (1981), Janssen & Bremer (2004), Pridgeon et al. (1997, 1999, 2001), Renz (1980), Schiestl et al. (1999), Schmid & Schmid (1977).

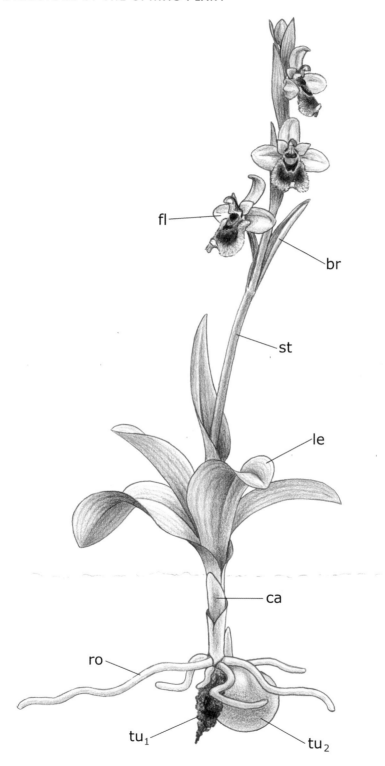

FIG. 1. *Ophrys tenthredinifera*. **tu₁**: old tuberoid; **tu₂**: young tuberoid; **ro**: root; **st**: stem; **ca**: cataphyll; **le**: foliage leaf; **br**: bract; **fl**: flower. Drawing by J. Lassen.

2. Structure of the *Ophrys* plant

Bee orchids (like all other orchids) are perennial herbs, the flowers of which have three sepals and three petals, in two closely-spaced whorls, and an inferior ovary. In the orchid flower, the sexual organs are united into a so-called column (gynostemium), while the seeds are minute, immensely numerous and devoid of endosperm. In common with the vast majority of other orchids, the ventral petal is differentiated into a so-called lip (labellum), the pollen grains are glued together to form pollinia, and only one anther (the median one of the outer whorl) is still well-developed and functioning – the others are vestigial (staminodes) or have vanished completely; moreover, the fruit is a capsule with one locule.

The following combination of features is held by *Ophrys* in common with the majority of other members of the *Orchis* subfamily: (1) the plant has underground tuberoids; (2) the pollinia are composed of smaller pollen masses, they are basally prolonged into stalks, and are attached to adhesive discs (viscidia) on either side of a prolonged median stigma lobe. The erect orientation of the anther places *Ophrys* in the *Orchis* tribe, while it agrees with the other genera of the *Orchis* subtribe with regard to the parallel anther locules and the fact that the fertile part of the stigma is placed in a cavity.

In preparation for a more definite description of the structure of the *Ophrys* plant, it is convenient to distinguish between vegetative and reproductive organs.

VEGETATIVE ORGANS

In bee orchids, the vegetative organs comprise root, stem and leaves as illustrated in FIG. 1. The structure of these parts is fairly constant throughout the genus and, therefore, contributes little information that can be utilised for the identification of species.

In principle, the root system is a fibrous root, but it is modified to consist mainly of two (rarely three) spherical to ovoidal tuberoids (i.e. "tubers" formed partly from the stem, partly from the root). From the point where the tuberoids are attached to the base of the stem proper, ordinary roots are produced. The internal tissue of the tuberoids consists of numerous small living cells that store nourishment in the shape of starch, and fewer, but larger, dead cells that store water and mucilage. These supplies enable the tuberoid to survive the adverse season and nourish the first developmental stages of the aerial shoot of the following year.

The stem is unbranched and usually 10-40 cm tall. In some species, however, flowering individuals of just c. 5 cm sometimes occur, whereas other species can produce giants with a stem of 90 cm or more. At the base of the stem, below ground level, a couple of pale leaf sheaths (so-called cataphylls) are situated. The fully differentiated leaves – the foliage leaves – are arranged in a rosette just above the ground. Each of them consists of a sheath that tightly embraces the stem and a spreading, curve-ribbed blade with a linear-oblanceolate to oblong-obovate outline. Sometimes, 1-2 foliage leaves are situated on the stem above the rosette. The foliage leaves are always uniformly (bluish) green. In the inflorescence, each flower is borne in the axil of a small, sheathless leaf called a bract. A few bract-like leaves are usually found between the uppermost foliage leaf and the inflorescence.

FLOWER, FRUIT AND SEED

The flowers as shown in FIG. 2. serve the sexual reproduction of the plant – and by appearing more or less insect-like they form part of the general adaptation of the genus to pollination by pseudocopulation (see From pollination to seed dispersal p. 24). The insect-like appearance is due to a combination of the individual shapes of several organs and varies considerably among species. Consequently, the flowers contain many important characters for the identification of species.

The sepals (the outer whorl of tepals, FIG. 2A-D) are of uniform size and always glabrous and more or less ovate. The colour ranges from white to (yellowish) green or pink, whilst the mid-vein is often distinct and green. In certain species, the lateral sepals can be bicoloured, so that the mid-vein forms a boundary between an upper and a lower half with separate colours. The petals (the lateral tepals of the inner whorl) are usually smaller than the sepals, sometimes tiny, and range in shape from triangular to oblong or linear. They are often hairy and can be wavy along the margin. They are, in general, more strongly coloured than the sepals, and in addition to the colours described above, one can encounter quite blackish-brown petals.

The lip (the median tepal of the inner whorl, FIG. 2A-H) is, in principle, a petal, but it is strongly modified to imitate the body of an insect. It is normally larger than the other tepals and ranges in outline from entire to deeply three-lobed, sometimes with the mid-lobe being retuse (shallowly notched) to emarginate (notched). The lip can end abruptly in a short point, or it can be provided with a larger, terminal appendage which is often rhomboid to obtriangular, dentate and erect. In many species, the

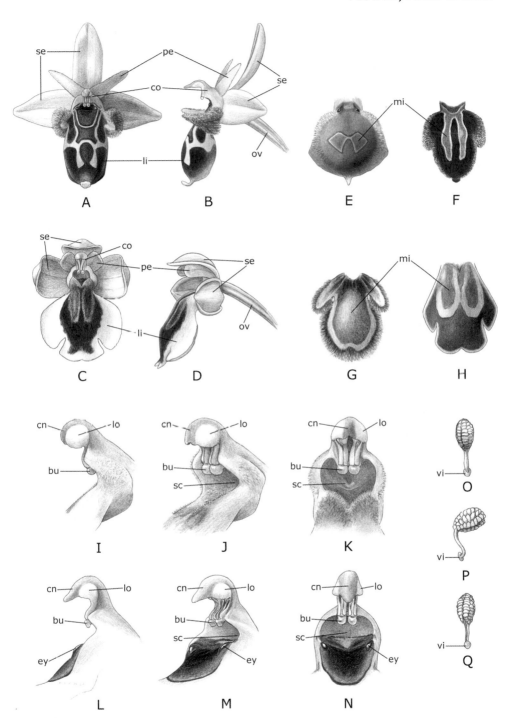

FIG. 2. **A-B.** Flowers of *Ophrys scolopax* subsp. *scolopax*, **C-D.** *O. lutea* subsp. *lutea*. **se**: sepal; **pe**: petal;
li: lip; **co**: column; **ov**: ovary. **E.** Lips of *O. argolica* subsp. *argolica* s.s., **F.** *O. sphegodes* subsp. *spruneri*,
G. *O. speculum* subsp. *speculum*, **H.** *O. fusca* subsp. *fusca*. **mi**: mirror. **I-K.** Columns of *O. fusca* subsp. *fusca*,
L-N. *O.* ×*arachnitiformis*. **cn**: connective; **lo**: anther locule; **bu**: bursicle; **sc**: stigmatic cavity; **ey**: eye-like knob.
O-Q. Pollinia of *O. fusca* subsp. *fusca*. **vi**: viscidium. Drawing by J. Lassen.

basal part of the lip exhibits two bulges that are most often conical, but can vary from minute humps to prominent, usually curved horns. In a three-lobed lip, the bulges can be formed from the side lobes, or they can be situated on the central part of the blade. In longitudinal section, the lip is in most cases slightly vaulted, but ranges from saddle-shaped to strongly vaulted. The margin can be anything from revolute to upcurved. Different parts of the upper surface are covered with one-celled hairs of different lengths and shapes. The ground colour of the upper lip surface is light to very dark brown, sometimes (brownish to greenish) yellow along the margin. Minute, dome-shaped papillae with smooth cell walls occur on the margin and on the distal part of the lower lip surface.

Nearly always, the centre of the upper lip surface is provided with a so-called mirror (speculum) of varying size (FIG. 2E-H). It can be square, H-shaped or more complicated; it can be coherent or fragmented; it can be diffuse or distinctly demarcated; it can be isolated or connected with the lateral edges or with the base of the lip. The mirror frequently shines with a greyish to bluish or reddish hue, reflecting ultraviolet rays and apparently imitating the wings of an insect. Though usually looking perfectly smooth to the human eye, the mirror is actually formed of minute one-celled trichomes with flattened bases and longitudinal cuticular striations. The expanded polygonal bases probably reflect most incident radiation. The cuticular striations, running from base to tip on each hair, probably increase the brightness of the mirror by scattering emergent light, regardless of the angle of incident light.

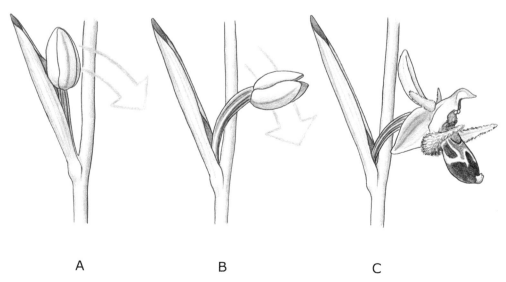

A   B   C

FIG. 3. Progression of resupination (stages **A-C**) in *Ophrys scolopax* subsp. *cornuta*. Drawing by J. Lassen.

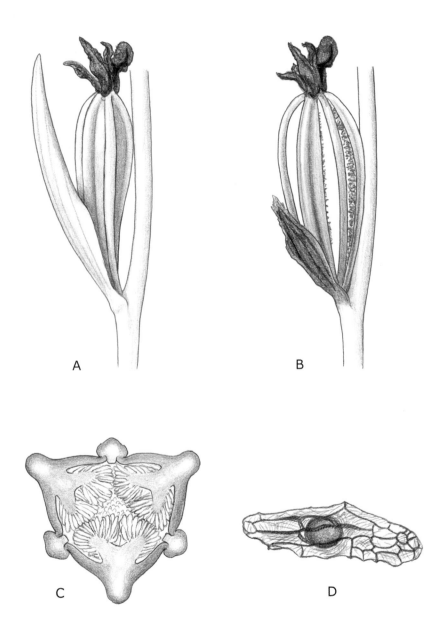

FIG. 4. Fruits and seed of *Ophrys sphegodes* subsp. *sphegodes*. A. Ripening capsule; B. dehisced capsule; C. transverse section of capsule; D. seed. Drawing by J. Lassen.

The column (FIG. 2I-N) is light green and roughly cylindrical with a somewhat widened base. In some species the tip of the column (actually, the sterile median part of the two-loculate anther: the connective) is extended and acute. Each locule of the anther contains a slender, stalked pollinium with a separate adhesive disc attached to its base (FIG. 2O-Q). The apical part of the extended median stigma lobe is sterile. It separates the lower parts of the anther locules and produces two tiny pouches (bursicles) full of liquid, each enclosing the adhesive disc of one pollinium. This prevents desiccation of the adhesive discs. The fertile part of the stigma is situated in a basal cavity (FIG. 2J-K, M-N) on the front side of the column. Two pale or dark, eye-like knobs are often found on the edge of this cavity.

The ovary (FIG. 2B,D) is light to darker green, sessile, fusiform-cylindrical and encloses a single locule. When the flower is still in bud, the lip is turned upwards (FIG. 3A), but when the flower is opening, it tips over (FIG. 3B), and the lip ends up pointing downward (FIG. 3C). *Ophrys* shares this phenomenon (termed resupination) with the majority of other orchids (but in most other orchids – and sometimes in O. *tenthredinifera* – the resupination is brought about by the pedicel and/or ovary twisting through 180° rather than by the flower tipping over). This means that an orchid flower appearing to be "correctly" orientated is, in fact, upside down! Upon fertilisation of the ovules, the ovary develops into a fruit in the shape of a capsule (FIG. 4A,C). When mature, the capsule dehisces by longitudinal slits (FIG. 4B).

The seeds are minute (0.3-0.6 mm long, 0.1-0.2 mm in diameter) and around 10,000-17,000 are usually produced in each fruit. The individual seed (FIG. 4D) consists of a few-celled embryo surrounded by the so-called carapace – a cell layer that protects the embryo against fungal and bacterial infections. The only reserves of the seed are bodies of fat and proteins contained within the cells of the embryo. The shielded embryo lies inside a very thin, inflated testa which is oblongoid and often somewhat curved; this inflated testa gives the seed a relatively large volume compared with its extremely low weight. The outer cell walls of the seed have secondary thickenings in the shape of transverse striations of the testa cells.

MONSTROUS FORMS

It is not unusual to encounter *Ophrys* individuals that are aberrant – most frequently with regard to the shape and/or colour of the sepals or petals, but sometimes with regard to the structure of the inflorescence or column. Among the most common monstrous (teratological) forms are individuals

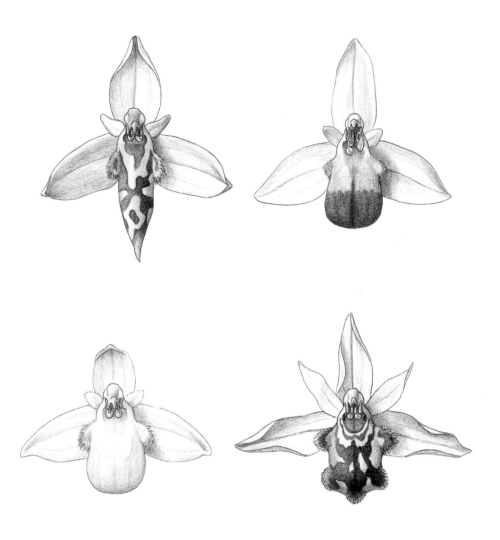

FIG. 5. Examples of monstrous forms of *Ophrys apifera*. These forms are often treated as distinct taxa, but in our opinion they do not deserve to be recognized at varietal level or above. Drawing by J.Lassen.

with flowers where the petals are shaped as the lip (or vice versa), or where the lip (and sometimes the column) is cleft into two. Also relatively frequent are plants with flowers in which the red and blue pigments (anthocyanins) are very poorly developed, and where the flower colour therefore is dominated by whitish, yellowish and greenish shades (hypochromic forms). Less common are individuals with flowers completely devoid of red and blue pigments (apochromic forms) or with flowers that contain extraordinarily high concentrations of these pigments and therefore are very strongly coloured in purplish and blackish–brown shades (hyperchromic forms).

It seems that monstrous forms can originate through developmental interruptions in one or more meristems (zones of cell-producing tissue) in a previously normal individual. In such spontaneous cases the aberrancy involves all or only some of the flowers and is typically present in one year only. Developmental interruptions can probably be induced by a number of factors (pollution, late frosts, etc.).

There is considerable evidence that aberrancies can also be transmitted through seeds produced by monstrous flowers. In such hereditary cases, the aberrancy will be exhibited in all flowers of the progeny throughout its lifetime. Monstrous forms are more frequent in the self-pollinated *Ophrys apifera* than in any of the insect-pollinated species (FIG. 5). Thus, self-pollination may imply that aberrant character states are more easily transmitted due to the lack of genetical recombination between individuals. Furthermore, it seems fair to assume that monstrous individuals of the self-pollinated species succeed in producing more progeny than monstrous individuals that depend on flowers with precise adaptations to a specific pollinator.

In addition to being exotic and fascinating, monstrous individuals representing heritable aberrancies ("hopeful monsters") have a potential to almost instantaneously generate new species – a phenomenon known as "saltational evolution". In most cases, however, they are nothing but transient forms. Still, they may give important cues in connection with developmental genetic research into the evolution of the orchid flower.

REFERENCES: Ascensão et al. (2005), Bateman & DiMichele (2002), Bateman & Rudall (2006), Baum et al. (2002), Dressler (1993), Johansen & Frederiksen (2002), Klein (1978), Kreutz (1997), Mrkvicka (1994), Nazarov & Gerlach (1997), Peitz (1967), Pridgeon et al. (2001), Rudall & Bateman (2002), Servettaz et al. (1994), Stern (1997), Szlachetko & Rutkowski (2000), Ziegenspeck (1936).

3. Biology, ecology and distribution

Like all other orchids, *Ophrys* exhibits several fascinating biological and ecological adaptations. As far as life cycle, germination and symbiosis with fungi are concerned, the bee orchids resemble most other members of the family. On the other hand, the adaptation to pollination by insects through pseudocopulation (see the section From pollination to seed dispersal p. 24) is exceptional. Otherwise, this phenomenon has only been observed in representatives of a few orchid genera from Australasia (*Arthrochilus, Caladenia, Chiloglottis, Cryptostylis, Drakaea, Leporella, Spiculaea,* and possibly *Caleana, Calochilus, Diuris, Lyperanthus, Pterostylis*), from India/Sri Lanka (*Cottonia*), and from tropical America (*Lepanthes, Mormolyca, Trichoceros, Trigonidium,* and possibly, *Stellilabium, Telipogon, Tolumnia*).

The life and annual cycles offer an appropriate point of departure for describing the biological and ecological features of bee orchids.

GERMINATION AND ADOLESCENCE

One thing that *Ophrys* has in common with all other orchids is the fact that no cotyledon develops during germination – only the so-called protocorm, which is a minute, tuber-like structure. The protocorm grows and produces tiny, scaly leaves whereupon the whole structure is called a mycorrhizome. In *Ophrys* the mycorrhizome has roughly the shape of an upturned onion. The scant reserves in the embryo cells can only nourish the very first stage of germination. This takes place at the beginning or middle of summer, after which the future of the plant depends entirely on acquisition of nourishment from the surroundings.

Sometimes the protocorm, possibly already the seed, will become infected by a fungus of the genus *Epulorhiza* – and only by digesting the microscopic hyphae of the fungus can the young *Ophrys* plant take up sufficient nutrients to survive and develop further in its natural environment. Symbiosis between a fungus on the one hand and the root (and frequently other underground organs) of a plant on the other is called mycorrhiza and is found in one form or another in most plant families. In some cases the mycorrhizal symbiosis involves mutual exchange of nutrients, but orchids seem to be pure parasites on their fungi.

During the first autumn the steadily growing mycorrhizome starts to produce roots, and by the beginning of the next summer it has developed a tiny, spherical tuber. Only this tuber survives the summer. During the second autumn, a short, thick rhizome shoots from the meristem at the apex of the tuber. The rhizome develops profusely mycorrhiza-forming roots and is terminated by a leafy aerial shoot. In early spring additional roots are produced that play a more important role in water absorption. Later during the spring another tuber develops from the base of the leafy shoot. The first flowering of the *Ophrys* plant usually takes place 1-7 years after the first aerial shoot appeared. The period required to reach this point seems to depend somewhat on the species – for example, *O. sphegodes* usually flowers in the first year that it appears above ground.

Under laboratory conditions with constant environment and no interspecific competition, the developmental rate and sequence appears to be slightly different (FIG. 108).

ANNUAL GROWTH CYCLE AND VEGETATIVE REPRODUCTION

Depending on the species and on the situation of the locality, *Ophrys* flowers some time within the period (January-)February-August (-September). In most parts of southern Europe peak flowering occurs during March-May. Each single flower can, in principle, stay fresh and open for more than three weeks, but it will wither about one week after being pollinated. In unfavourable environmental conditions, flowering may not take place for one or more years, and only a rosette of foliage leaves appears above ground. The plant can even live completely below ground for up to two years (perhaps longer). The relative proportion of flowering individuals in an *Ophrys* population in one particular year depends not on the local environmental conditions alone, but also on the species in question.

At the time the flowers are opening, the leaves begin to wither (FIG. 6A). When the flowering, fruit ripening and seed dispersal (FIG. 6B) are over (i.e. in late spring or summer according to the geographic situation), the remainder of the plant dies down, the only exceptions being a fresh tuberoid and the still hidden bud predestined to become the next year's aerial shoot. Now the plant survives virtually as an underground tuberoid for one or more months. This adaptation effectively counteracts injurious desiccation during mid- and late summer which, especially in the Mediterranean, is characterised by lack of water.

Both the aerial and underground organs of the *Ophrys* plant are renewed annually. The new tuberoid begins to develop little by little in the autumn (FIG. 6C-D) and can be recognised throughout the winter as a small excrescence on the shoot just above the old tuberoid. Also, the new roots

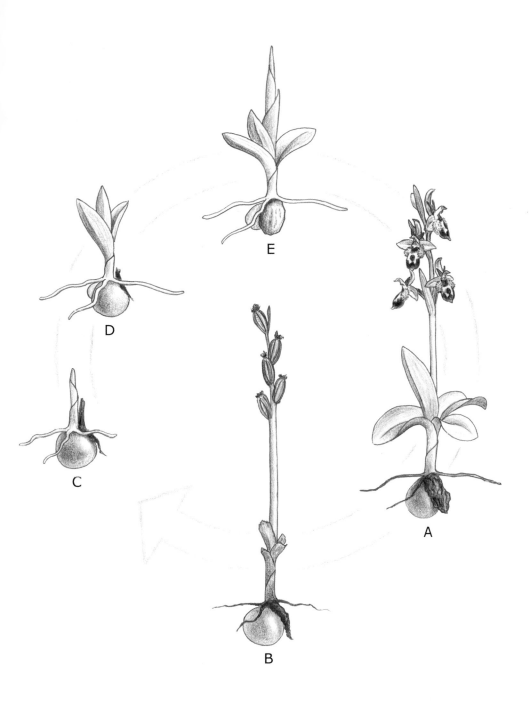

FIG. 6. Annual growth cycle of *Ophrys reinholdii*. **A.** late spring; **B.** summer; **C.** early autumn; **D.** late autumn; **E.** early spring. Drawing by J. Lassen.

start to appear, and they are immediately infected by the mycorrhizal fungus which has survived the dry summer season in the outer cell layers of the *Ophrys* tuberoid. In subtropical areas the new rosette of foliage leaves is likewise produced during the autumn (FIG. 6D). In temperate areas, on the other hand, the bee orchids are not consistently winter-green – many individuals do not form their new leaf rosette until early spring.

The coming of spring involves a marked increase in the growth of the new tuberoid (FIG. 6E). Assisted by the photosynthesis of the plant, the tuberoid now builds a reserve of nourishment for the next dry hibernation period and subsequent sprouting. At the same time the old tuberoid shrinks, while its reserves (mainly starch) are being utilised for the formation of the flowering shoot. At the end of the cycle, the old tuberoid has virtually disappeared. Thus, an *Ophrys* individual at the beginning of flowering normally has one young, well-developed, fresh tuberoid and one old, shrunken tuberoid (FIG. 6A).

Under favourable conditions of growth, vegetative reproduction can take place by the old tuber being replaced by more than a single new one. Usually, vegetative reproduction plays a very modest role. However, it is of considerable importance in *O. bombyliflora* which is also peculiar in its new tuberoids being produced at the end of stolons several centimetres long. Therefore, efficient vegetative reproduction results in this species forming large groups of flowering shoots that appear 5-10 cm from one another. In all other *Ophrys* species, shoots produced by vegetative reproduction will be densely clustered.

FROM POLLINATION TO SEED DISPERSAL

The most important method of reproduction in all species of *Ophrys* is sexual; this can be roughly divided into pollination, fertilisation, fruit development and seed dispersal.

Pollination is the process by which pollen is transferred to a receptive stigma. As in nearly all other orchid genera, *Ophrys* is basically adapted to pollination by insects. When an appropriately sized insect lands on the flower and places itself in a suitable position on the lip, either its head or abdomen will bump against the two liquid-containing pouches (bursicles) on the sterile stigma lobe. In this way, the insect comes into contact with the adhesive discs that effectively attach the stalked pollinia to the head or abdomen of the insect.

Insect-pollinated *Ophrys* species have adaptations that enhance the relative probability of outbreeding. Immediately after the pollinia have been

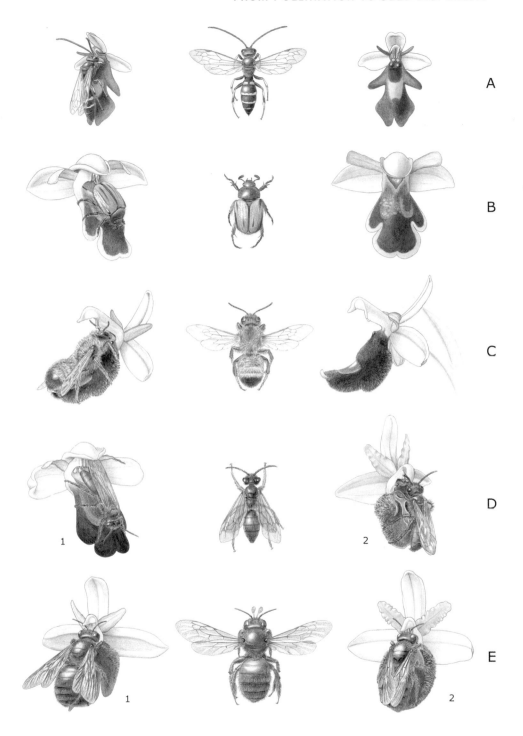

Fig. 7. Examples of *Ophrys* flowers and their pollinators. **A.** *Argogorytes mystaceus/Ophrys insectifera* subsp. *insectifera*; **B.** *Blitopertha lineolata/Ophrys fusca* subsp. *blitopertha*; **C.** *Chalicodoma parietina/Ophrys bertolonii*; **D.** *Andrena morio/Ophrys fusca* subsp. *iricolor* (**1**)/*O. sphegodes* subsp. *atrata* (**2**); **E.** *Xylocopa iris/Ophrys sphegodes* subsp. *spruneri* (**1**)/*O. sphegodes* subsp. *sipontensis* (**2**). Drawing by J. Lassen.

removed from the anther they are erect and, therefore, do not come into contact with the fertile stigma part below. Because of unequal desiccation the pollinia stalks now bend gradually forwards, resulting in a lowering of the pollinia. However, several minutes elapse before the pollinia reach an angle fitting the receptive part of the stigma, and by this time the insect has usually left to visit a flower on another individual. Consequently, it will frequently be a flower on this second individual that is pollinated.

Ophrys is pollinated almost exclusively by solitary bees (of the families Andrenidae, Anthophoridae, Apidae, Colletidae, Megachilidae and Xylocopidae). The only exceptions known are *O. speculum* subsp. *speculum* and *O. insectifera* subsp. *insectifera* (FIG. 7A), both pollinated by solitary wasps (of the families Scoliidae, Sphecidae and Argidae), and *O. fusca* subsp. *blitopertha* (FIG. 7B) which is unique in being consistently pollinated by a beetle, more specifically a chafer (family Scarabaeidae).

In *Ophrys* it is always male insects that visit and pollinate the flowers. The minute dome-shaped papillae on the lip margin and on the distal part of the lower lip surface emit a scent that can hardly be perceived by the human nose. Nevertheless, the scent can attract males of particular insect species. Indeed, the male insects perceive the scent so efficiently that they can even find experimentally hidden flowers. The chemical composition of the scent is usually very complex (appropriately referred to as an "odour bouquet") and imitates sex pheromones emitted from the cuticle of females of the pollinator species. Experiments with *O. sphegodes* and its pollinator, conducted by F. P. Schiestl and co-workers, suggest that alkanes and alkenes together constitute the sex pheromones of the female bees as well as the

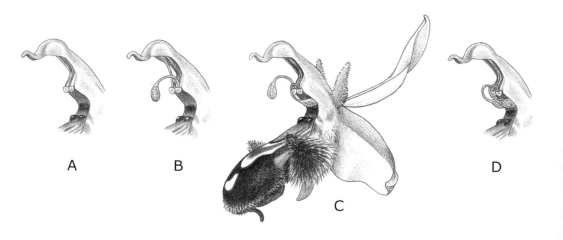

A   B   C   D

FIG. 8. Progression of self-pollination (stages **A-D**) in *Ophrys apifera*. Drawing by J. Lassen.

floral odour component inducing mating behaviour of the male bees (see below). The specific attractants in the odour bouquet of the flower appear to be identical with those of the female bees. In contrast to *O. sphegodes*, the olfactory attraction to flowers of *O. speculum* operates with a simple system of only a few compounds.

Composition of the floral odour bouquet varies considerably, in particular with regard to the biologically non-active compounds. Variation is encountered between flowers in the same inflorescence, between individuals and between populations. It has been observed that male bees recognise and avoid flowers they have visited earlier. However, as they willingly visit other flowers either of the same or a different individual, this phenomenon may increase the reproductive success of the orchid.

Ophrys flowers do not offer nectar or any other reward to visiting male insects. On the other hand, it is obvious that the flowers also morphologically imitate the females of their pollinators (FIG. 7). The insect-like appearance is mainly due to the more or less hairy lip which furthermore has a markedly three-dimensional conformation and a usually shining mirror (FIG. 2E-H). The latter reflects ultraviolet light and imitates the play in folded insect wings. Two eye-like knobs (in some species merely spots), frequently seen on the edge of the stigmatic cavity, likewise deserve mention (FIG. 2L-N). They have sometimes been interpreted as eye-imitations, but according to H. F. Paulus they more likely imitate the scale-like structures (tegulae) that overlap the bases of the fore wings in bees and wasps. Finally, it should be noticed that the petals are sometimes very narrow and reminiscent of insect antennae – this effect is especially striking in *O. insectifera* (FIG. 7A).

The insect-like facies (appearance) of the *Ophrys* flower may contribute to short-distance attraction of potential pollinators. However, there is hardly any doubt that it plays a more important role in stimulating the male insect to assume the "correct" position on the lip – the specific position that can lead to foreign pollinia being deposited on the stigma and to the flower's own pollinia being removed from the anther. In this context it should be specially noted that the hairiness of the lip very precisely imitates the hairiness found on the back of the female insect. In *O. atlantica*, *O. fusca*, *O. lutea* and *O. omegaifera*, all of which lack eye-like knobs on the edge of the stigmatic cavity, the pollinator positions itself with the abdomen turned towards the column (FIG. 7B,D1). In all other species of *Ophrys* it positions itself with the head towards the column (FIG. 7A,C,D2-E). In a few cases, a pollinator is shared between two *Ophrys* species without causing frequent hybridisation, simply because it assumes opposite positions on the flowers of those species (FIG. 7D). It seems that the direction of the hairs of the lip is of particular importance in inducing the pollinator to turn either its abdomen or its head towards the column. The former takes place in species with hairs pointing towards the column, the latter in species with hairs pointing away from the column. In species with a terminal appendage on the lip the pollinator always positions itself with the head towards the column, and it evidently takes the appendage for a female abdominal apex. In short, it seems that the male insect mistakes the flower for a related female ready for mating, and sometimes several males compete for the same flower.

Having assumed a suitable position on the flower, the male attempts to copulate with the lip. It is in connection with this "pseudocopulation" that the pollinator often deposits and/or removes pollinia. The insect does not pull its punches, and it is sometimes observed that the lip is even perforated by its penis. A pseudocopulation event may last up to 15 minutes, but normally takes only a few seconds. It appears usually to be broken off without ejaculation of semen – probably due to the belated recognition by the male insect of crucial differences between the real female insect and the lip of the flower. Therefore it is also possible that the sexual motivation of the male is maintained after the first visit to a flower, and if this is the case, it increases the chances of an immediate visit to another flower, which again increases the chances of (cross-)pollination.

In a recent study of *O. sphegodes*, F. P. Schiestl and M. Ayasse observed a significant increase in the amount of all-*trans*-farnesyl hexanoate in the floral odour bouquet after pollination and interpreted this as a signal to guide pollinators to unpollinated flowers of the inflorescence. Farnesyl hexanoate is a major constituent of the Dufour's gland secretion in females of the pollinator bees. It is used in the lining of brood cells, but it also induces males to reduce the frequency of copulation attempts. The composition of the floral odour bouquet also changes in other ways upon pollination, and a significant reduction in the total amount of scent emitted takes place during the following two to four days.

The special type of insect pollination seen in *Ophrys* can be regarded as parasitism on the sexual behaviour of the pollinators. The reason for the relative success of the deceit is probably that the males of the pollinator species normally appear (up to four weeks) earlier than the females and that the *Ophrys* plants mainly flower in the period where only males are present. Although unpollinated *Ophrys* flowers will continue to attract and excite males long after females have emerged, it seems likely that pollination would be less efficient under more competitive conditions.

It is a vexed question whether the insect pollination is species-specific (i.e. whether each single *Ophrys* species is always pollinated by only one species of insect). This problem is dealt with more explicitly in the chapter Evolution, hybridisation and classification p. 44.

Ophrys apifera deviates from other *Ophrys* species in being self-pollinating (although there is evidence of occasional pollination by male bees). When the flower opens, the curved anther continues its growth (FIG. 8A), and in this way the pollinia are gradually pulled out of the locules (FIG. 8B). Only the basal parts of the pollinia, together with the adhesive discs, remain uninfluenced. Possibly due to internal cavities, the stalks of the pollinia are flaccid, for which reason the pollinia tilt forwards and downwards (FIG. 8C). Aided by gusts of wind, the thick pollen-bearing part of each pollinium subsequently pivots directly onto the receptive part of the stigma (FIG. 8D).

Spontaneous selfing is generally less advantageous than pollination by insects. This is mainly due to the fact that the consistent inbreeding following self-fertilisation reduces the genetic variation of the species and, consequently, its adaptability. Furthermore, it increases the risk of genetically determined deaths. The evolution of those adaptations that in many plants (e.g. all *Ophrys* species except *O. apifera*) increase the probability of cross-pollination, as compared with selfing, has undoubtedly taken place against this background. Nevertheless, selfing also offers certain advantages – chiefly the assurance of a consistently high rate of seed production. In plants having a genetically fortunate constitution, the advantages of selfing may amply counterbalance its disadvantages, and it is in this light that the evolution of

FIG. 9. Characteristic *Ophrys* habitats – maquis (Italy, Toscana, Monte Argentario, 14th April 2001). Habitat of *O. argolica* subsp. *crabronifera*, *O. lutea* subsp. *lutea* and *O. sphegodes* subsp. *litigiosa* var. *argentaria*. Photo by N. Faurholdt.

spontaneous selfing in O. *apifera* should be seen.

When one or more pollinia have been deposited on a receptive stigma, each of the many thousand pollen grains will germinate with a so-called pollen tube. The pollen tubes grow from the stigma down through the style component of the column and eventually reach the ovarian cavity where thousands of ovules are situated side by side in longitudinal files. Here the egg cell in each of the ovules will be receptive to a single pollen tube. When the pollen tube has established a direct contact with the egg cell, fertilisation takes place by the transfer of immobile male gametes through the pollen tube (a phenomenon known as siphonogamy). The exact length of time separating pollination and fertilisation in *Ophrys* is unknown – but usually about two weeks will pass after pollination, until an incipient

swelling of the ovary indicates that fertilisation has probably taken place.

Fertilisation represents the beginning of seed development (the final structure of the seed is briefly described in the section Flower, fruit and seed p. 14). As the seeds grow, the ovary swells significantly, and towards the end of seed development it dries up into a brown capsule. In early summer or mid-summer the capsule dehisces by longitudinal slits (FIG. 4B), and the seeds are subsequently dispersed by the wind. A considerable share of the minute seeds undoubtedly fall to the ground fairly close to the mother plant, but there is no theoretical limit to the distance that the seeds can float in the air. It therefore seems likely that the geographic distributions of individual *Ophrys* species are limited by ecological parameters (relating to germination, dormancy, pollination etc.) rather than restrictions on seed dispersal.

FIG. 10. Characteristic *Ophrys* habitats – maquis (Italy, Sicily, Bosco di Ficuzza, 20th April 1999). Habitat of *O. bertolonii*, *O. bombyliflora*, *O. fusca* subsp. *pallida*, *O. lutea* subsp. *galilaea*, *O. lutea* subsp. *lutea* and *O. tenthredinifera*. Photo by H. Æ. Pedersen.

LIFE-HISTORY STRATEGIES

Up to now, detailed studies of life-history strategies in the genus have been conducted on *O. apifera* and *O. sphegodes* only. However, the strategies of these two species appear strikingly different.

To date, the most thorough demographic studies on *O. apifera* are those conducted by T. C. E. Wells and R. Cox at Monks Wood, England. They found that typically about 25% of the non-dormant individuals flower in a given year. Additionally, the number of flowering individuals fluctuates strongly from one year to another, apparently depending on weather conditions. In *O. apifera*, the new tuberoid, bound to nourish the formation of the next flowering shoot, is not formed until very late in the autumn. For this reason, its further development (and, hence, the probability of flowering next spring) is strongly influenced by the temperature and precipitation patterns of the winter in question. The generally low frequency of flowering in *O. apifera* is counterbalanced by the longevity of the

Fig. 11. Characteristic *Ophrys* habitats – maquis (France, Provence, Pont de Grenouillet, 3rd May 1990). Habitat of *O. scolopax* subsp. *scolopax* and *O. sphegodes* subsp. *sphegodes*. Photo by H. Æ. Pedersen.

Fig. 12. Characteristic *Ophrys* habitats – garrigue (Spain, Balearic Islands, Mallorca, S'Arenal, 17th April 1990). Habitat of *O. bombyliflora, O. fusca* subsp. *fusca, O. omegaifera* subsp. *dyris, O. speculum* subsp. *speculum, O. tenthredinifera, O. ×brigittae* and *O. ×flavicans.* Photo by N. Faurholdt.

individual plants. Thus, an individual typically lives about 7 years (quite frequently more than 10) after it first appears above ground, and it is likely to flower repeatedly during its lifetime. A few studies conducted elsewhere generally support the findings of Wells and Cox, although there is some evidence that the flowering frequency may in some populations be higher and the longevity of individual plants shorter.

In a population of *O. sphegodes* in Sussex, M. J. Hutchings has found that typically more than 80% of the non-dormant individuals flower in a given year. Additionally, the number of flowering individuals is relatively constant from one year to another. More than 70% of the individuals already flower in the first year that they appear above ground. On the other hand, many of them die upon their first flowering, and only few individuals live for more than three years.

The strategic differences between *O. apifera* and *O. sphegodes* are striking. The individual plants of the former can live for a long time and get several opportunities to reproduce, notwithstanding the fact that they do not flower each year. Furthermore, the self-pollination in this species will ensure a consistently high seed production. In *O. sphegodes*, on the other hand, individual plants cannot be expected to live for a long time. Reproduction is secured through the high frequency of flowering in this species and through the early development of flowering shoots.

It would be tempting to interpret the strategic differences as adaptations to different types of habitat – *O. apifera* being adapted to stable environments, *O. sphegodes* to less stable ones. Nevertheless, *O. apifera* is the one behaving most like a pioneer species. It is frequently seen as one of the earliest colonisers of bare ground in abandoned quarries, on railway embankments and roadside verges etc., but it generally disappears when the vegetation develops a closed canopy. *Ophrys sphegodes*, on the other hand, mainly belongs in more stable environments (though occasionally forming vast populations in ruderal habitats, for instance on spoil, from the channel tunnel, dumped on either side of the English Channel). This discrepancy between expected and observed habitat preferences obviously calls for further studies.

OPHRYS HABITATS

The suitability of a local area as habitat for one or more *Ophrys* species depends on whether the area meets all the ecological requirements of bee orchids in connection with their germination, adolescence, annual growth cycle and pollination. Firstly, the climate as well as the structure, nutrient content and moisture of the soil must be reasonably close to the optima for *Ophrys* – otherwise the bee orchids cannot compete successfully with associated plants. Secondly, the area has to meet a number of more specific needs which imply the presence of appropriate mycorrhiza-forming fungi and pollinators in the shape of particular insect species.

In general, the above requirements are best complied with in light-

Fig. 13. Characteristic *Ophrys* habitats – garrigue (Italy, Sicily, Necropoli di Pantalica, 18th April 1999). Habitat of *O. bertolonii*, *O. fuciflora* subsp. *biancae*, *O. fuciflora* subsp. *oxyrrhynchos*, *O. fusca* subsp. *fusca*, *O. lunulata*, *O. lutea* subsp. *galilaea*, *O. lutea* subsp. *lutea*, *O. speculum* subsp. *speculum*, *O. sphegodes* subsp. *atrata*, *O. sphegodes* subsp. *sphegodes* and *O. tenthredinifera*. Photo by H. Æ. Pedersen.

FIG. 14. Characteristic *Ophrys* habitats – garrigue (Greece, Attika, Imettos near Athens, 30th March 1996). Habitat of *O. scolopax* subsp. *cornuta*, *O. ferrum-equinum* subsp. *ferrum-equinum*, *O. fusca* subsp. *fusca*, *O. lutea* subsp. *galilaea*, *O. lutea* subsp. *melena*, *O. speculum* subsp. *speculum*, *O. sphegodes* subsp. *aesculapii*, *O. tenthredinifera* and *O. umbilicata* subsp. *umbilicata*. Photo by N. Faurholdt.

open, relatively warm and dry areas with fairly nutrient-poor, calcareous (neutral to basic) soil. However, certain species of *Ophrys* sometimes occur in relatively acid or moist habitats. Habitats especially preferred by the individual species and subspecies are indicated in the systematic account. In this section, only a general survey will be given.

The vast majority of *Ophrys* species are found in the Mediterranean – in areas with subtropical, Mediterranean climate (i.e. warm, dry summers and mild, humid winters), as well as in areas on the boundary between Mediterranean and temperate climate (characterised by cooler summers, colder winters, and precipitation that is more evenly distributed over the year). Therefore, it is reasonable to begin the habitat tour in these areas. Most interest centres on the various vegetation types derived from the original forest. These have originated through long-term human exploitation of the evergreen oak forest, which is now very sparse but formerly covered extensive areas in the Mediterranean. Excessive felling,

Fig. 15. Characteristic *Ophrys* habitats – garrigue (Greece, Naxos, Halkio, 16th April 2000). Habitat of *O. fuciflora* subsp. *andria*, *O. lutea* subsp. *lutea* and *O. reinholdii*. Photo by N. Faurholdt.

burning and grazing have in most places degraded the oak forest to shrub communities – according to the height of the scrub known as maquis (c. 1-4 m) or garrigue (up to c. 1 m). "Maquis" and "garrigue" are French words, but "macchia" (Italian), "barocal" (Portuguese), "tomillares" (Spanish), "gariga" (Italian), and "phrygana" (Greek), may also be encountered.

The shady oak forest supports very few of the generally light-loving bee orchids. Almost the same is true for the maquis (FIGS 9-11), mainly composed of drought-resistant evergreen trees and shrubs, often furnished with spines, poison, or a repulsive taste as protection against grazing animals. The most essential species belong to the genera *Pistacia* (family Anacardiaceae); *Ruscus* (Asparagaceae); *Arbutus*, *Erica* (Ericaceae); *Quercus* (Fagaceae); *Phillyrea* (Oleaceae); *Paliurus* and *Rhamnus* (Rhamnaceae). In this kind of scrub, however, there are often glades that may accommodate fine *Ophrys* populations.

Bee orchids are predominantly found in the low, light-open garrigue (FIGS 12-15), and at propitious sites they can grow in profusion with several species and subspecies occurring in mixed colonies. Among the shrubs and subshrubs dominating such places, representatives of the following genera are particularly characteristic: *Cistus* (family Cistaceae); *Euphorbia*

(Euphorbiaceae); *Calycotome, Cytisus, Genista* (Fabaceae); *Lavandula, Rosmarinus, Teucrium* (Lamiaceae) and *Sarcopoterium* (Rosaceae). Like the trees and shrubs of the maquis – perhaps even more obviously – the dominating plants of the garrigue are thoroughly adapted to surviving under conditions of summer-drought, grazing and frequent fires. In addition to *Ophrys*, the garrigue accommodates numerous other tuberous and bulbous herbs that similarly escape burning and damaging desiccation by passing the summer in dormancy below ground. Throughout the spring, from March to May, the flowering of these herbs is a grand spectacle. However, even garrigue can suffer excessive grazing, typically resulting in a so-called pseudo-steppe, dominated by asphodels and generally lacking in species; few bee orchids grow in such impoverished places.

Lowland pine forest (composed of *Pinus halepensis*, *P. pinea*, and/or *P. pinaster*) is at present considerably more abundant than evergreen oak forest in the Mediterranean. Where the pine forest is open (FIG. 16) and grows in suitable soil it may hold very fine *Ophrys* populations. However, it is difficult to predict where to find these excellent patches – one can often walk for a long time through exceedingly drab pine forest and suddenly

FIG. 16. Characteristic *Ophrys* habitats – pine forest (Greece, Rhodes, Attaviros, 17th April 1996). Habitat of *Ophrys fuciflora* subsp. *candica*, *O. fuciflora* subsp. *fuciflora*, *O. fusca* subsp. *fusca*, *O. lutea* subsp. *galilaea*, *O. omegaifera* subsp. *omegaifera* and *O. reinholdii*. Photo by H. Æ. Pedersen.

encounter a first-class *Ophrys* site.

Among the most prominent *Ophrys* habitats in the Mediterranean we also find roadsides with low, species-rich vegetation (FIGS 17-18), and abandoned vineyards and open, predominantly old, olive groves that are run without application of fertiliser, pesticides and mechanical treatment of the soil (FIG. 19).

Further north in Europe both the Mediterranean vegetation types and the great majority of *Ophrys* species disappear. Hardly any bee orchids but *O. fuciflora*, *O. sphegodes*, *O. apifera* and *O. insectifera* occur in the truly temperate parts of Europe. In northern Europe, the bee orchids are even more dependent on calcareous soil than they are in the south, and all of the above species can be found on chalk grassland (FIG. 20). In particular *O. apifera* also occurs on stabilised coastal dunes (FIG. 21), occasionally even on the exposed sand dunes nearest the sea. *Ophrys insectifera* often grows in calcareous fens, typically characterised by species of *Schoenus* (family Cyperaceae), or on light-open ground in pine forest, scrubs, wooded meadows (FIG. 22) and edges of woods – rarely in shady beechwoods.

The bee orchids are mainly lowland plants, as most of the species only occur up to c. 1400 m altitude (and only two reach 2000 m). The climatic

changes that occur with increasing altitude are in many respects similar to those that occur with increasing latitude, and the changes in vegetation types are in many ways parallel. As might be expected therefore, those *Ophrys* species that in the Mediterranean occur at higher altitudes frequently grow on grassland as well as in open scrub and edges of woods, which are often reminiscent of *Ophrys* habitats in temperate lowland areas.

DISTRIBUTION PATTERNS

The European range of the genus *Ophrys* is indicated on MAP 1. The total range additionally includes parts of North Africa (northern areas in Morocco, Algeria, Tunisia and Libya), Cyprus, the Middle East, the Caucasus and Anatolia, as well as other parts of the Near Orient – east to the southern end of the Caspian Sea and south to the northern end of the Persian Gulf. Very few species of *Ophrys* are absent from Europe, though at the subspecific level the figure is somewhat higher.

Among the 19 *Ophrys* species occurring in Europe, *O. insectifera* (MAP 10) is restricted to temperate areas, whereas *O. apifera* (MAP 11), *O. sphegodes* (MAP 18) and *O. fuciflora* (MAP 14) are found in both temperate and

(left) **FIG. 17.** Characteristic *Ophrys* habitats – roadsides (Greece, Naxos, Moni, 19th April 2000). Habitat of *O. reinholdii* and *O. scolopax* subsp. *scolopax*. Photo by N. Faurholdt.

(right) **FIG. 18.** *Ophrys sphegodes* subsp. *litigiosa* var. *argentaria* growing on roadside (Italy, Toscana, Monte Argentario, 6th April 2001). Photo by N. Faurholdt.

subtropical regions. The remaining 15 species generally belong in subtropical areas. *O. fusca* (MAP 5) and *O. lutea* (MAP 6), however, extend far north along the French Atlantic coast, and *O. scolopax* (MAP 13) extends to the southern fringes of the East European steppes.

The 15 species that (almost) exclusively occur in subtropical areas represent different distribution types within the European part of the range – mainly differentiated along an axis in an east-west direction. Six species – *O. tenthredinifera* (MAP 7), *O. speculum* (MAP 9), *O. fusca* (MAP 5), *O. lutea* (MAP 6), *O. scolopax* (MAP 13) and *O. bombyliflora* (MAP 8) – are distributed all the way from east to west (only the last, however, reaches the Canary Islands). *Ophrys atlantica* (MAP 3) is only found in a small area in southernmost Spain (in immediate connection with its main range in North Africa), whereas *O. bertolonii* (MAP 17) and *O. lunulata* (MAP 19) are only known from the central part of the Mediterranean. *Ophrys argolica* (MAP 15) occurs in a narrow zone from Italy, across the Peloponnese, to the Aegean Islands, and the four following species are even more pronouncedly eastern, as they are only distributed from mainland Greece eastwards: *O. kotschyi*

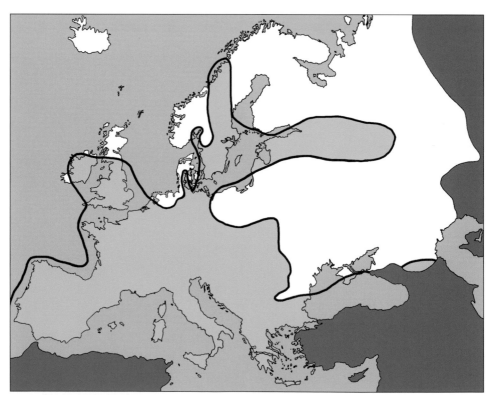

MAP 1. The distribution of genus *Ophrys* in Europe, where 19 species are known to occur. The European country richest in bee orchids is Greece with 17 species; only *O. atlantica* and *O. lunulata* are missing.

Fig. 19. Characteristic *Ophrys* habitats – olive groves (Greece, Lesbos, Moria, 31st March 1996). Habitat of *O. ferrum-equinum* subsp. *ferrum-equinum*, *O. fusca* subsp. *fusca*, *O. fusca* subsp. *iricolor*, *O. lutea* subsp. *galilaea*, *O. scolopax* subsp. *scolopax* var. *minutula*, *O. speculum* subsp. *speculum* and *O. tenthredinifera*. Photo by N. Faurholdt.

(Map 21), *O. reinholdii* (Map 20), *O. ferrum-equinum* (Map 16) and *O. umbilicata* (Map 12). Finally, it should be mentioned that *O. omegaifera* (Map 4) has a particularly interesting, divided (disjunct) distribution. Thus, it is represented (by various subspecies) in three widely separated areas – a western area, a central one, and one in the eastern part of the Mediterranean.

The number of species (the specific diversity) decreases towards the corners of the European range. Only *O. insectifera* (Map 10) reaches Norway and Sweden as well as the easternmost parts of Europe north of the Black Sea, and only *O. apifera* (Map 11) reaches the northern part of Ireland, whereas only *O. bombyliflora* (Map 8) and *O. sphegodes* (Map 18) reach the Canary Islands and the lowland north of the Caucasus, respectively. Furthermore, it should be noticed that only *O. insectifera* has been capable of colonising the central parts of the Alps. At the opposite end of the spectrum of diversity it is striking that the subtropical Mediterranean accommodates the great majority of *Ophrys* species.

Diversity at the subspecific level shows the same geographic tendencies

FIG. 20. Characteristic *Ophrys* habitats – chalk grassland (England, Kent, Dover, 17th July 1993). Habitat of *O. sphegodes* subsp. *sphegodes*. Photo by H. Æ. Pedersen.

as the species diversity. The importance of the Mediterranean, however, is even more evident. Likewise, it is particularly characteristic at the subspecific level that by far the most marked diversity centres of *Ophrys* within Europe are in Greece and southern Italy.

As mentioned above, *Ophrys* encompasses widely as well as narrowly distributed species. Among the former we find especially *O. insectifera* (MAP 10), *O. sphegodes* (MAP 18) and *O. apifera* (MAP 11), each of which occupies more than half of the total range of the genus. At the opposite end of the

FIG. 21. Characteristic *Ophrys* habitats – stabilized coastal dunes (Wales, Merioneth, Morfa Harlech, 14th June 1999). Habitat of *O. apifera*. Photo by H. Æ. Pedersen.

Fig. 22. Characteristic *Ophrys* habitats – calcareous wooded meadow (Denmark Zealand, Allindelille Fredskov, 12th June 1993). Habitat of *O. insectifera* subsp. *insectifera*. Photo by H. Æ. Pedersen.

spectrum we find a number of very narrowly distributed – "endemic" – species. Among these, *O. lunulata* has a particularly narrow distribution, apparently being confined to Sicily.

Subspecies similarly exhibit wide to narrow distributions – but in any case, subspecies are generally more narrowly distributed than species. Thus it is hardly surprising that endemics are mostly found at the subspecific level. Some of the endemics have extremely narrow ranges, examples being *O. sphegodes* subsp. *sipontensis* (known only from the Siponte plateau of Monte Gargano in southern Italy) and *O. fuciflora* subsp. *andria* (known only from four Cycladean islands in the Aegean Sea).

Endemics at specific as well as subspecific level are of considerable interest for the scientific study of evolution. Since, due to their narrow distributions, they are generally vulnerable, they also commonly receive great attention in conservational contexts.

REFERENCES: Ågren et al. (1984), Ascensão et al. (2005), Ayasse et al. (2000, 2003), Baumann & Baumann (1990), Bjørndalen (2006), Blanco & Barboza (2005), Bockhacker (1996), Borg-Karlson (1990), van der Cingel (1995, 2001), Claessens & Kleynen (2002), Dafni & Bernhardt (1990), Edmondson (1979), Hermjakob (1976), Hill (1978), Hutchings (1987, 1987a, 1989), Kullenberg (1956, 1961, 1973a), Kullenberg & Bergström (1976), Kullenberg, Büel & Tkalců (1984), Möller (1989, 1996, 2000), Neiland & Wilcock (1995), Nelson (1962), Paulus (1998, 2006), Paulus & Gack (1980, 1990), Pridgeon et al. (2001), Priesner (1973), Rasmussen (1995), Sanger & Waite (1998), Schick & Seack (1988), Schiestl & Ayasse (2001), Schiestl et al. (1997, 1999, 2000), Singer (2002), Singer et al. (2004), Stahl (1989, 1993), Stoutamire (1974, 1975), Summerhayes (1951), Sundermann (1961, 1962, 1962a, 1962b), Warncke & Kullenberg (1984), Wells & Cox (1989, 1991), Wiefelspütz (1964), Wolff (1950, 1951).

4. Evolution, hybridisation and classification

On many occasions the genus *Ophrys* has been described as a rapidly evolving plant group. As outlined below, there have been good reasons for this. It has also been claimed that all the many subspecies, in the course of time, will evolve into species (even if species are defined according to the unusually broad concept applied in the present book). However, we do not share the latter view; subspecies and varieties generally seem to be relatively short-lived entities that come and go.

Hybridisation is a phenomenon that is most commonly noticed through the occurrence of fortuitous first-generation hybrids, the possible progeny of which are unable to compete. In special cases, however, recurrent hybridisation and backcrossing to the parental species can lead to blurring of species boundaries on the one hand or to the start of new speciation processes on the other.

Ongoing evolution, as well as the various effects of hybridisation within *Ophrys*, present several almost inextricable problems to the efforts of appropriately classifying the species and the entities below species level. Many authors have tried to solve these problems in a lot of different ways, and we have had to deal with the same difficulties ourselves. For this reason we find it natural to discuss evolution, hybridisation and classification together in a single chapter.

DIVERGENT EVOLUTION

New evolutionary entities with a potential to become new species originate when barriers preventing (or at least heavily restricting) genetic exchange with other populations of the same species are present. A reproductive isolation of this kind can arise, for example, when a species is represented by populations on different islands or when different (sub)populations of a species (by individual adaptations through mutations) develop a dependency on different specific pollinators or types of habitat. Events of chromosome-doubling seem to have been an important factor for the origin of the tetraploid *O. fusca* and *O. omegaifera*; all other *Ophrys* species with known chromosome numbers are normally functionally diploid. When two or more isolation mechanisms occur together they may usefully reinforce the effect of one another.

When various (sub)populations of a species are more or less reproductively isolated from each other they will often evolve in different directions. Such divergent evolution can happen by two kinds of natural selection. So-called directional selection leads to divergence of geographically distinct populations, whereas disruptive selection leads to divergent evolution within a single colony. In both cases, individuals with a particular composition of genes (i.e. a particular genotype) produce more viable and fertile progeny than individuals with other genotypes. Consequently, in the course of generations, they secure their own genes a higher frequency within the population in question. In *Ophrys*, the reason why individuals with a particular genotype are favoured like this might be, for example, (1) that the successful genotype involves a better adaptation to climatic extremes in the area, (2) that it involves a flowering time that is closer to the optimum as defined by the phenology of the pollinator, or (3) that it codes for an odour bouquet that more precisely matches the female sex pheromones of the pollinator. Each of these phenomena would mean a strengthening of the competitive power, and it stands to reason that natural selection in mutually isolated (sub)populations of one and the same species will frequently result in divergent evolution. As an alternative to directional or disruptive selection the divergent evolution can take place through accumulation of fortuitous genetic changes over a series of generations. Fortuitous changes like these will, statistically, move in different directions in mutually isolated populations – a phenomenon called genetic drift. Small population size tends to increase the rate of divergent evolution.

Provided that divergent evolution within a species passes unimpeded it will lead to the formation of new species or, at least, new subspecies. During this process the accumulating genetic differences will normally cause an increasing intersterility between the populations involved. In certain groups, like birds and mammals, the point of completion of a speciation process is usually obvious, but as far as flowering plants (and many other groups) are concerned, it is a much debated question. Definitions of the categories species, subspecies and variety within *Ophrys* are discussed more explicitly in the section The classification developed for this book p. 55.

In the light of pollination studies it is assumed that divergent evolution in *Ophrys* often happens by directional or disruptive selection that optimises the adaptations of different (sub)populations to different species of pollinators. It has been suggested that the relative amounts of alkenes in the floral odour bouquet are particularly responsible for selective attraction of specific pollinators in *Ophrys*. Since attracting a new pollinator species may only require a change in odour pattern (not necessarily synthesis of new compounds), divergent evolution in *Ophrys* is likely to be a rapid process.

The European distribution patterns of the genus (see the section Distribution patterns p. 39) indicate that geographic barriers must also play an important role for the evolution in *Ophrys*. This is particularly evident from the high diversity of both species and subspecies in Greece – a country that offers numerous opportunities for mutual geographic isolation of populations in connection with the islands in the Aegean Sea, as well as the mountains of the mainland.

It is less clear to what extent different flowering times constitute a significant parameter influencing divergent evolution in *Ophrys*. For a species that is pollinated by a range of insect species, some populations may experience relative mutual densities of pollinator species that are different from those at other sites. Under such conditions, variation in flowering time may enhance divergent evolution through gradual adaptation of local populations to phenologically differing pollinators dominating the individual sites.

A considerable number of those entities that are in this book recognised as subspecies, as well as a few species, appear to be pollinated (normally) by only one insect species each. It is interesting that these highly specialised subspecies are more narrowly distributed than their specific pollinators – unexplained by climate, dispersal biology or available types of habitat. This overall pattern suggests that the subspecies concerned are young entities that have simply had insufficient time for expanding to all parts of their potential distribution areas. Our hypothesis that the present fine-scale diversity (i.e. below species level) in *Ophrys* results from a high systematic turnover (or very recent radiation) is supported by DNA data recently published by M. Soliva, R. M. Bateman and their research groups.

Because the above pattern recurs in a high proportion of the subspecies in the genus it is tempting to suppose that subspecies in *Ophrys* are often relatively short-lived – they may arise and die out at regular intervals. Possible reasons for recurrent extinction of subspecies could include drastic population reductions (permanent or temporary) in the specific pollinator. Another explanation could be that previously separated *Ophrys* populations with a shared pollinator species expand, come into contact and finally merge through a prolonged process of repeated hybridisation and backcrossing (introgression).

HYBRIDS AND HYBRID COMPLEXES

Natural hybridisation occurs frequently in *Ophrys* – between species as well as between subspecies or varieties within a species. With regard to features such as structure and colouration patterns, hybrids are usually more or less intermediate between the parents (FIG. 23). In general, therefore, the more different the parents look, the easier it is to detect and identify a hybrid, and hybrids between different species are thus much easier to recognise than hybrids between subspecies or varieties within a species. In any case, the increasing ease of DNA sequencing should soon remove much of the uncertainty and anecdotal nature of current attempts to identify hybrids. Certain molecular data in combination even have the power to reveal which species is mother and which is father of a given hybrid.

Fortuitous hybrids (FIG. 24) are not rare among bee orchids, and hybrids between subspecies or varieties of the same species appear to be as fertile as their parents. In most cases, however, they are less competitive. The reason might be that the parents are generally depending on different

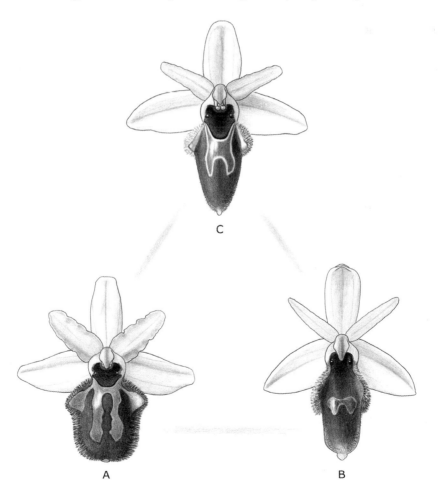

FIG. 23. A. *Ophrys sphegodes* subsp. *atrata*; B. *O. lunulata*; and C. a fortuitous first-generation hybrid between them. Drawing by J. Lassen.

47

FIG. 24. Fortuitous *Ophrys* hybrids. **A.** *O. fuciflora* subsp. *apulica* × subsp. *oxyrrhynchos* (Italy, Puglia, Martina Franca, 20th April 2002); **B.** *O. argolica* subsp. *biscutella* × *tenthredinifera* (Italy, Puglia, Monte Gargano, 11th April 1995); **C.** *O. fuciflora* subsp. *andria* × *scolopax* subsp. *scolopax* (Greece, Naxos, Koronos, 18th April 2000); **D.** *O. bombyliflora* × *speculum* subsp. *speculum* (Portugal, the Algarve, Malhao, 31st March 1999); **E.** *O. fuciflora* subsp. *oxyrrhynchos* × subsp. *parvimaculata* (Italy, Puglia, Martina Franca, 20th April 2002). Photos by N. Faurholdt.

FIG. 24B

FIG. 24C

FIG. 24D

FIG. 24E

specific pollinators or habitats. In both cases the adaptations of the intermediary hybrid will usually be inferior compared with those of either parent, for which reason the hybrid will be less able to compete. Nevertheless, the fertility of the hybrid can under certain conditions bring about either a merger of previously distinct entities or a new incident of divergent evolution.

In the section Divergent evolution p. 44 brief mention was made of the possibility that extensive genetic exchange through hybridisation and backcrossing could, in the long term, result in a merger of different bee orchids with shared pollinators. This might happen if different subspecies, following long–term divergent evolution in geographically separate areas, come into contact while expanding to new regions. There is little doubt

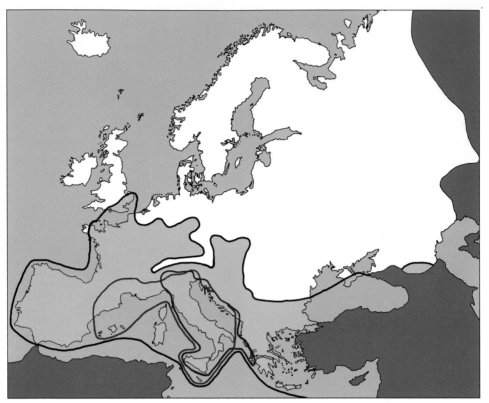

MAP 2. The distribution of the partly stabilised hybrid complex *Ophrys bertolonii × sphegodes* (red), the distribution of *O. bertolonii* (blue) and the European distribution of *O. sphegodes* (black).

that this phenomenon occurs occasionally within *Ophrys*, but no concrete examples have been established so far. One possible case (in an advanced stage) is that of *O. umbilicata* subsp. *umbilicata*. This subspecies exhibits a clinal variation from Albania in the northwestern corner of the total range, to Iran in the southeast: plants with green sepals reign supreme (or very nearly so) in southeastern Europe, but they are gradually replaced by plants with rose-coloured to white sepals that constitute by far the dominant form at the opposite end of the range. One explanation, though not the only possible one, is that range expansions have resulted in a geographic overlap between two previously separate populations (a western one exclusively consisting of plants with green sepals, and an eastern one exclusively consisting of plants with rose-coloured to white sepals), and that a genetic merger of the two populations is currently occurring through hybridisation and backcrossing.

Hybridisation can also be effective in the diametrically opposite direction by triggering a new incident of divergent evolution. This most obviously happens when the hybrid turns out to be not only fertile but also competitive in a certain area where it starts to evolve in another direction than either of the parental species. However, genetic drift may also play an important role.

Already the first-generation hybrid might accidentally produce a floral odour bouquet which attracts a pollinator species different from those of the parents. In that case there will be a strong selection pressure towards an optimisation of the pheromone imitation concerned (and probably other adaptations), meaning that the hybrid evolves decisively away from both parental species in the course of generations.

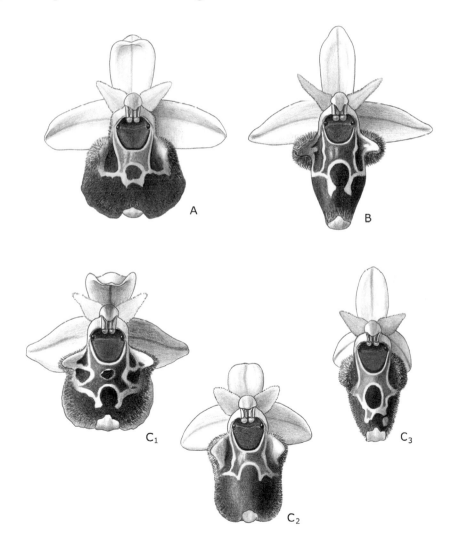

FIG. 25. **A.** *Ophrys fuciflora* subsp. *fuciflora*; **B.** *O. scolopax* subsp. *scolopax*; and **C$_{1-3}$.** representative forms of their partly stabilised hybrid complex, *O.* ×*vicina* nm. "*heterochila*". Drawing by J. Lassen.

Another phenomenon appears more commonplace, wherein the fertile hybrid proves to be competitive in a certain area, even though it employs the same pollinator as at least one of its parents. Such fertile and competitive hybrid complexes may even occur between subspecies of different species (exactly which subspecies are involved cannot always be positively deduced from the morphology of the hybrid). The interspecific hybrids *O. fusca* × *omegaifera*, *O. argolica* × *scolopax*, *O. bertolonii* × *sphegodes*, *O. fuciflora* × *sphegodes*, and *O. fuciflora* × *scolopax* are all represented by partly stabilised hybrid complexes. The fertility of the hybrid complexes and their present identity as more or less self-contained evolutionary entities can be seen from the fact that the geographic range of each hybrid complex only partially overlaps with the area where both parental species occur. For instance, the partly stabilised hybrid complex *O. bertolonii* × *sphegodes* occurs on both Malta and the Balearic Islands as well as in southern France – areas where *O. bertolonii* is absent (MAP 2). At the same time, however, the lack of their own particular features and the often extremely high level of variation within each hybrid complex (FIG. 25) strongly suggest continued recurrent backcrossing to one or both parental species, meaning that the hybrid complexes have not yet formed completely stable and isolated entities.

DIFFERENT KINDS OF CLASSIFICATIONS

Systematic classifications of bee orchids are typically based on comparisons of structural differences and similarities and/or comparative assessments of mutual relationships or reproductive isolation/interaction.

Structural differences and similarities can be compared in various ways, for example in the shape of a herbarium revision where appropriate species boundaries are established through the sorting of numerous herbarium specimens. The establishment of these boundaries can be based either on estimates or statistics. Although the herbarium revision is a suitable method for systematic studies of many other groups of plants, it must generally be considered unfit for the classification of bee orchids. Firstly, many of the important characters of these plants are lost in the drying process. Secondly, it is first and foremost the complicated variation at population level that is the crucial point when boundaries between species as well as subspecies and varieties are going to be drawn. Such patterns of variation in *Ophrys* are often impossible to decipher by a herbarium revision – the real biological basis for statistical testing is uncertain as the dried specimens available from various herbaria have not been collected according to a standardized sampling method in the natural populations.

Several recent classifications of *Ophrys* mainly rely on findings from older herbarium revisions, but additionally include many species that were later described (often separately) on the basis of what could be referred to as "striking new finds" (for example, a botanist with a reasonable field experience might have stumbled upon one or more *Ophrys* populations that proved impossible to identify using the existing literature). One of the problems associated with descriptions based on "striking new finds" is the combination of a slender data set and a rather intuitive and premature systematic decision.

Studies of morphological variation within and between natural populations, often incorporating multivariate statistical methods, have constituted the basis of many systematic alterations in *Ophrys* in recent years. Such studies contribute useful knowledge of variation patterns within the genus – usually the discussion is focused on statistically significant inter-population differences in the mean values of selected characters. However, with regard to the utilisation of the results for formal classifications, a high proportion of the published population studies of *Ophrys* suffer from a substantial deficiency: they do not incorporate tests that clarify to which extent individual characters can reliably distinguish between the tentative species when a single individual must be identified. Furthermore, the following circumstances provide food for thought: (1) the individuals sampled from each study population rarely seem to be chosen according to a standardised method that would secure a reasonable amount of objectivity; (2) frequently, only quantitative characters are included; and (3) it is often assumed that all characters exhibiting significant differences between populations are genetically fixed. Thus, the possibility of environmentally governed morphological plasticity is commonly ignored, and influence from the developmental stage of the individual plant measured and the position of the sampled flower on the spike is largely disregarded. Additionally, S. Malmgren has reported that: "The method of large-scale propagation also makes it possible to study the very wide range of variation in [*Ophrys*] plants grown from a single seed capsule, even when obtained from a self-pollinated plant. Such variation is seen in wild populations too and accounts for many of the differences that have been used to describe "new" species and subspecies". In conclusion, we think that statistical population studies have a great potential for the classification of *Ophrys*, but we also think that the currently prevalent methods are in need of modification and standardisation before the studies can lead to widely applicable conclusions.

Probable interrelationships among bee orchids can be hypothesised by so-called cladistic analyses that produce theoretical phylogenetic trees on the basis of certain principles. One of the most important principles is the assumption that the shortest phylogenetic tree based on available data, that is the tree that involves the lowest possible number of evolutionary events, constitutes the most probable phylogenetic hypothesis. Cladistic analyses are normally based on morphological characters, on molecular data (usually from DNA sequences) or on a combination of the two. Cladistics conducted on morphological characters have occasionally been applied to *Ophrys*, but usually simply in order to uncover the probable relationships between already accepted species. The same is true for cladistic analyses conducted on molecular data. In the latter case, however, the still sparse results are interesting because some of them suggest the need for a wider species concept than those that prevail in recent accounts on *Ophrys*. Additional results of molecular studies on *Ophrys* are eagerly awaited!

The extent to which populations or groups of populations are reproductively isolated from each other can be studied in several ways. Real quantitative analyses of hybridisation and backcrossing within *Ophrys* are rather few. However, such studies form the substantial basis of the treatment of partly stabilised hybrid complexe in this book.

Far more frequently, the line of approach to the study of reproductive isolation in *Ophrys* has been to examine the importance of pollinator specificity as an isolation mechanism, and H. F. Paulus and C. Gack in particular, are responsible for many contributions in this field. They have employed a special and interesting method, the idea of which is that the researchers seek out the nests of a particular pollinator species and place a selection of flowering *Ophrys* plants in the immediate vicinity. In this way it soon becomes clear which individuals the particular pollinator species is attracted to and which it ignores. The results suggest that *Ophrys* species should be defined extremely narrowly due to a generally very high and consistent pollinator specificity. However, a lot of the recently described species represent the extremities of a more or less gradual and continuous spectrum of variation (in characteristics such as morphological characters and flowering time), and in many cases it seems that no individuals from the central part of this spectrum have been tested in the pollination study. Whenever the response of the pollinators to the succession of intermediary forms is unknown, it is impossible to know whether the extreme forms, apparently characterised by different pollinator specificity, are really reproductively isolated from each other.

Other authors, including the great pioneer in the study of *Ophrys* pollination, B. Kullenberg, have generally found pollination in *Ophrys* to be less species-specific, and this is further supported by the frequent instances of hybridisation between even morphologically well-defined species. The conflicting hypotheses concerning the degree of pollinator-specificity in *Ophrys* should obviously be tested by assessing patterns and levels of gene flow in mixed colonies.

Regardless of being based on intuitive, population statistical or pollination biological data, the major part of the systematic work conducted on *Ophrys* in recent years show two distinct trends. Firstly, the categories of subspecies and variety are very often ignored – all accepted entities are recognised as species. Secondly, a strong, progressive splitting at the species level is taking place, with ever more, ever narrower species being described. As will appear from the following section, we are sceptical of the validity of both trends.

THE CLASSIFICATION DEVELOPED FOR THIS BOOK

As long as they are within the limits of the *International Code of Botanical Nomenclature*, all classifications of *Ophrys* are equally valid, no matter on which criteria they are based and no matter whether a narrow or wide species concept is applied. However, this neutral point of departure does not prevent considerable debate regarding the appropriateness of conflicting *Ophrys* classifications.

The whole idea of formal classification and naming of plants is to define and designate recognisable entities to which reference can be made in connection with (popular as well as scientific) communication about the appearance, relationships, biology, ecology and conservation of the plant etc. This necessarily implies that a classification, to be fully operational, must allow not only populations but also individual plants to be reliably identified to species level (and, as a rule of thumb, we believe that a success rate of at least 90% should be demanded). Perhaps a slightly higher degree of uncertainty should be accepted in connection with identification of an individual to subspecies or variety, because subspecies and varieties are often considered less stable than species. Even at the subspecific and varietal levels, however, we think that if an individual cannot be identified with at least 85% certainty (as a rule of thumb), then the classification is not useful. At the same time we want to stress that we are not claiming that all less morphologically well-defined entities are uninteresting and without biological relevance. On the contrary, such vaguely defined forms (FIG. 26) probably indicate instances of recently initiated divergent evolution. Consequently, they are exceedingly interesting and should be studied in detail – but to assign them the rank of species, subspecies or just variety seems unhelpful.

The currently prevailing trends (1) to ignore the categories of subspecies and variety, and (2) to assign species rank to even the smallest groups of populations that are recognised in a given classification, we consider inexpedient for three reasons. Firstly, this practice renounces the obvious possibility of letting the classification and scientific plant names reflect different levels of, for example, morphological similarity, phylogenetic relationship and/or reproductive isolation. Secondly, it implies that far too many individuals cannot be identified with certainty to species level. Thirdly, the pronounced splitting at the species level can easily lead to circular reasoning in connection with interpretation of distribution patterns and pollinator specificity (e.g. the circular logic of: "all known *Ophrys* species are pollinated by only one insect species each" ⇔ "if a population turns out to be pollinated by another insect species, it must represent a new species, since each species in the genus is known to be pollinated by only one insect species").

Aiming at the perfect classification of *Ophrys* (admittedly a Utopian goal!) it would be useful if, in the foreseeable future, definitions of the categories species, subspecies and variety were established on the basis of empirical observations on the nature of this particular genus. Similar definitions have been proposed for the likewise complicated orchid genus *Dactylorhiza* (the marsh orchids and spotted orchids). The latter definitions integrate apparent phylogenetic relationships (inferred from genetic data) as well as observations on the morphological variation and estimates of reproductive isolation between the entities (inferred from comparative statistical population studies). A corresponding basis for classification of *Ophrys* has not yet been provided. Therefore, we have had to let the classification in this book rest on a more theoretical basis. Our taxonomic criteria are broadly in accordance with H. Sundermann's views as repeatedly expressed by him during the period 1964-1987. More precisely, we have employed the following definitions:

1. A *species* consists of all individuals that under natural conditions (in reality or potentially) can interbreed to produce consistently viable and fully fertile offspring. Furthermore, the individuals of one species are distinguished from those of other species by morphological features.

2. A *subspecies* consists of a subset of populations of a species that differ morphologically from other subsets and, furthermore, in reality is

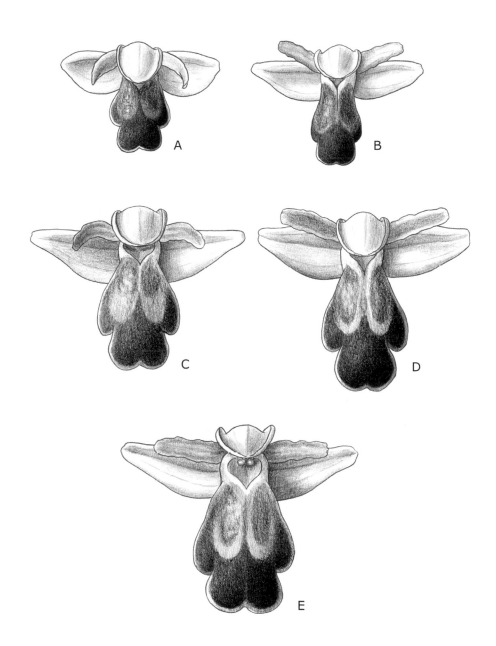

FIG. 26. Examples of slightly different forms of *Ophrys fusca* subsp. *fusca*. These forms appear to have specific pollinators, (form **A.**) viz. *Andrena tomora*; (form **B.**) *A. wilkella*; (form **C.**) *A. merula*?; (form **D.**) *Colletes cunicularius*; (form **E.**) *Andrena flavipes*. Drawing by J. Lassen.

reproductively isolated from these by one or more barriers prior to pollination. The latter barriers can be geographical (different distributions), phenological (different flowering times) or ecological (adaptations to different habitats or pollinators).

3. A *variety* consists of a subset of populations of a (sub)species that differ morphologically from other subsets, but without exhibiting obvious barriers that can ensure reproductive isolation from the latter. Contrary to *forms* (which we have chosen not to treat explicitly in the systematic account), varieties usually do not occur as stray individuals but tend to form local, often mutually exclusive, populations.

Some support for our practical delimitation of species according to the species definition above comes from largely unpublished data on experimentally produced *Ophrys* hybrids (S. Malmgren, in litt.). Thus, the following first-generation hybrids (indicated as maternal × paternal parent) produced no viable seed when they were self-pollinated: *O. apifera* × *fuciflora*, *O. apifera* × *scolopax*, *O. fuciflora* × *lutea* (and the reciprocal cross), *O. insectifera* × *fuciflora*, *O. insectifera* × *lutea*, *O. insectifera* × *speculum*, *O. insectifera* × *tenthredinifera*, *O. lutea* × *speculum*, *O. lutea* × *tenthredinifera*. Self-pollinated first-generation plants of *O. fuciflora* × *kotschyi* subsp. *cretica* (and the reciprocal cross) produced only small amounts of viable seed, and the progeny was very weak and unable to compete. The very low, or entirely missing, production of viable seed in the above hybrids seems mainly due to pollen sterility; it is unlikely that genetic self-incompatibility is involved, as self-pollination of non-hybrid *Ophrys* flowers usually results in copious seed set (S. Malmgren, pers. comm.).

In general, the practical application of the above definitions in the present book suffers to some extent from inadequate genetic and biological data. Consequently, the current classification should not be regarded as our ultimate idea of the most appropriate classification of the European bee orchids. Nevertheless, we believe that the system presented in this book offers a reasonable starting point for a more scientific attempt to construct a fully reliable and operational *Ophrys* classification.

A particularly serious problem we have had to deal with is the partly stabilised hybrid complexes that occasionally exist between entities that in all other respects behave as mutually distinct species (see the section Hybrids and hybrid complexes p.46). We basically consider the hybrid complexes as comparable to subspecies, as far as stability and evolutionary status are concerned. However, they cannot be classified as subspecies since, in that case, they should be placed and named under both parental species. On the other hand, we feel that it would be anomalous to attribute species rank to entities that are vaguely delimited towards the parental species, only partly stabilised and have a questionable evolutionary future.

After careful deliberation we propose that partly stabilised hybrid complexes between species should, as a general rule, be referred to under their binary hybrid names: *O. ×brigittae* for *O. fusca × omegaifera*, *O. ×delphinensis* for *O. argolica × scolopax*, *O. ×flavicans* for *O. bertolonii × sphegodes*, *O. ×arachnitiformis* for *O. fuciflora × sphegodes*, and *O. ×vicina* for *O. fuciflora × scolopax*. We think it would be advantageous if this practice of reserving the use of binary hybrid names for partly stabilised hybrid complexes could gain a general footing in connection with *Ophrys*. Even at best, however, it cannot become more than a generally accepted practice, as any opposing (or ignorant) person can legitimately continue to apply the binary hybrid names to quite fortuitous first-generation hybrids, too, without violating the *International Code of Botanical Nomenclature*.

Although we consider the classification in this book a step forward, we do not pretend that we have been able to deal with all of the many challenges in a satisfactory way. One concrete case may serve as example. Particularly in Campania, southern Italy, a bewildering jumble of forms can be observed in *Ophrys argolica* subsp. *biscutella*. On Monte Alburni, and in adjoining areas to the south, many populations are dominated by individuals with *O. fuciflora*-like petals and lip appendages and with a mirror that is usually connected to the base of the lip. We think that extensive hybridisation and backcrossing between *O. argolica* subsp. *biscutella* and *O. fuciflora* subsp. *fuciflora* is taking place, but we also hypothesise that the putative hybrid complex is insufficiently stabilised to deserve systematic recognition. However, we are uncertain whether this assessment is correct, just as we question the validity of our decisions in several other especially tricky cases. Under all circumstances, we are convinced that the most useful solution to such complicated problems will not be a continued splitting of the bee orchids into still narrower species, but rather accumulation of more rigorous data.

REFERENCES: Arnold (1997), Bateman (1999, 2001), Bateman & DiMichele (2002), Bateman et al. (2003), Bateman & Rudall (2006), Bernardos et al. (2003, 2005), Biel (2002), Borg-Karlson (1990), Caporali et al. (2001), E. Danesch & O. Danesch (1972), O. Danesch & E. Danesch (1976), Del Prete (1984), Devillers & Devillers-Terschuren (1994), Ehrendorfer (1980), Gölz & Reinhard (1975, 1980), Grant (1981), Greilhuber & Ehrendorfer (1975), Grünanger et al. (1998), Gulyás et al. (2005), Kullenberg (1961), Levin (1993, 2000, 2001), Malmgren (2006), Paulus (2006), Paulus & Gack (1986, 1990), Pedersen (1998), Pedersen & Faurholdt (1997, 2002), Rieseberg et al. (2004), Rudall & Bateman (2002), Schiestl & Ayasse (2002), Soliva et al. (2001), van Steenis (1957), Sundermann (1964, 1972, 1975, 1976, 1987), Vöth & Ehrendorfer (1976).

5. Systematic account

This chapter serves two purposes – firstly, to constitute an operational tool for the identification of species, subspecies, varieties and partly stabilised hybrid complexes; secondly, to present illustrations and more specific information on each single systematic entity. The chapter opens with a generic description, succeeded by a dichotomous key to the species and partly stabilised hybrid complexes, followed by accounts of each accepted species (1–19) and the hybrid complexes (H1–5).

The first key (including notes) should, in principle, make it possible to identify any *Ophrys* individual to species or partly stabilised hybrid complex. It should be added that applying the key to individuals reasonably typical of the population will, in general, facilitate the identification process. Furthermore, it is always a good idea to consult descriptions and illustrations while using the key. Finally, it should be noted that the key takes into account neither fortuitous first-generation hybrids nor monstrous forms, and consequently, beginners may sometimes experience problems when using the key; this difficulty will diminish as field experience increases.

Each species account contains a description of the plant, a discussion of its variation patterns and information concerning its flowering season, distribution and habitat requirements. For all species that are represented by more than one subspecies in Europe, the species account includes a key to the subspecies in the area.

The accounts of subspecies immediately follow the account of the species to which they belong. These accounts contain the same categories of information as the species accounts, but are generally shorter and do not repeat information that applies to the entire species. The accounts of varieties are composed and positioned according to the same principles.

At all systematic levels, each account contains an indication of the accepted name, after which is listed a number of synonyms (if any). This list does not pretend to be complete. On the contrary, it only contains names that can be encountered as accepted names of European entities in one or more of the orchid floras and dissertations marked with asterisks in the Bibliography and references section p. 272 (in the section Dubious records p. 226, synonyms include alternative names under which the plants have been reported from Europe). The objective is to make clear the relation between the many names and systematic status of the bee orchids accepted in those widespread publications on the one hand, and in this book on the other. Via the publications cited, it should be a simple matter to compose more complete lists of synonyms.

Likewise at all levels, each characterisation of distribution includes a list of the European regions from where the (sub)species, variety or hybrid complex is known. Due to the fact that several systematic entities are distributed far beyond Europe, we have found it appropriate also to indicate any occurrences at popular tourist destinations outside Europe, viz. Morocco, Tunisia, Anatolia, Cyprus and Israel, and, these, where included, are found at the end of each list. However, it should be borne in mind that the latter areas additionally accommodate subspecies (and a few species) that are not included in the present book. Abbreviations of geographic names p. 296 and delimitation of regions are explained at the end of the book.

Distribution maps are provided for all species and partly stabilised hybrid complexes. Each distribution map is based on all known occurrences of the bee orchid in question. We have preferred this model partly because it can be difficult to decide whether a species has disappeared completely from a given region, partly because regional (contrary to local) extirpation of bee orchids is still fairly limited. The ranges outlined on most of the maps give fairly reliable pictures of the present-day distributions.

REFERENCES: Baumann & Künkele (1982, 1986, 1988), Bournérias (1998), Buttler (1986), E. Danesch & O. Danesch (1969), Davies et al. (1983), Del Prete & Tosi (1988), Delforge (1994, 2001, 2005), Landwehr (1977), Mossberg & Nilsson (1987), Nelson (1962), Rossi (2002), Souche (2004), Sundermann (1980).

OPHRYS

Ophrys L. **Plant** compact to slender, 5-80(-90) cm tall with a basal rosette of linear-oblanceolate to ovate-oblong foliage leaves, 1-2 foliage leaves higher up on the stem and usually a few bract-like leaves below the inflorescence. **Spike** lax to dense with 1-15(-21) alternate (occasionally spirally arranged) flowers with ovate to linear-lanceolate bracts. **Sepals** white to purplish-violet, brownish, or (yellowish) green (sometimes bicoloured with the midvein as dividing line), (lanceolate-)oblong to (ob)ovate or broadly elliptic, 6-20 × 2-11 mm, glabrous. **Petals** whitish to purplish-violet, brownish, yellow or green, occasionally bright ruby, broadly triangular to linear with flat to wavy margins, sometimes auricled, 1-14 × 0.8-7 mm, glabrous to shaggy. **Lip** with (reddish) brown to (yellowish) green, grey or nearly black ground colour and sometimes a narrow to very broad yellow to green, light brown or reddish-brown margin, sessile or with a slender, stalk-like base, straight to saddle-shaped or gradually to abruptly downcurved, spread out or with recurved sides or margins, sometimes longitudinally furrowed at base, entire to deeply three-lobed, 5-28.5 × 5-27(-30) mm, with strongly varying hairiness (often concentrated along the margin), frequently provided with two hardly recognisable to long, horn-shaped bulges; front edge deeply cleft to obtuse, often ending abruptly in a short point or provided with a more conspicuous appendage; the central and/or basal part of the lip provided with a dull to shining, differently coloured mirror, in outline varying from extensive, complex patterns to small, isolated spots and dashes. **Column** rounded to acute; the fertile part of the stigma situated in a cavity below the anther and the erect sterile part of the stigma; the stigmatic cavity sometimes provided with lateral, eye-like knobs at base; the median anther fertile (the lateral ones vanished or vestigial), erect with two parallel locules containing separate pollinia; pollinia composed of numerous small masses of pollen, basally extended into stalks and joined with adhesive discs enclosed in liquid-containing pouches on either side of the erect sterile stigma lobe.

KEY TO THE SPECIES AND PARTLY STABILISED HYBRID COMPLEXES

1. Column rounded 2
1. Column acute (to obtuse) 10

2. Stigmatic cavity approximately as wide as the anther 3
2. Stigmatic cavity approximately twice as wide as the anther 4

3. Dorsal sepal distinctly boat-shaped, strongly incurved. Petals elliptic to ovate-triangular. Lip conspicuously shaggy along the margin, otherwise glabrous; mirror almost completely covering the mid-lobe . . . . 7. ***O. speculum***
3. Dorsal sepal nearly flat, straight. Petals linear. Lip velvety except on the mirror which covers less than half of the mid-lobe . . . . 8. ***O. insectifera***

4. Dorsal sepal more or less reflexed. Petals hairy. Lip with a terminal appendage or short point (sometimes hidden underneath the lip) 5
4. Dorsal sepal from the base more or less parallel to the column. Petals glabrous. Lip devoid of a terminal appendage or short point 7

5. Sepals (yellowish) green. Mirror obscure . . . . 6. ***O. bombyliflora***
5. Sepals violet to white with a green mid-vein. Mirror distinct 6

6. Lip 9-16 mm long, with a prominent tuft of hairs close to the tip . . . . 5. ***O. tenthredinifera***

6. Lip 7–9 mm long, without a prominent tuft of hairs close to
    the tip . . . . H5. **O. ×vicina** (nm. "*heterochila*")

7. Lip with spreading or upcurved margin . . . . 4. **O. lutea**
7. Lip with recurved margin 8

8. Petals recurved, with wavy margins. Lip saddle-shaped with
    a slender, stalk-like base; mid-lobe hardly longer than the
    side lobes . . . . 1. **O. atlantica**
8. Petals spreading to porrect (stretched forwards), with (almost)
    flat margins. Lip straight to downcurved, sessile; mid-lobe distinctly
    longer than the side lobes 9

9. Lip with a distinct longitudinal furrow at the base; mirror
    not ending in a distinct ω-shaped band (though often more
    brightly coloured in front) . . . . 3. **O. fusca** [but see note 1]
9. Lip (very nearly) devoid of a longitudinal furrow at the base;
    mirror ending in a distinct (in subsp. *hayekii* often obscure)
    white to blue, ω-shaped band . . . .
    . . . . 2. **O. omegaifera** [but see note 2]

10. Column extended in an S-curved tip; pollinia with flaccid
    stalks . . . . 9. **O. apifera**
10. Column acute (to obtuse), not extended; pollinia with firm
    stalks 11

11. Lip terminally with (rarely without) a small triangular to
    awl-shaped (very rarely almost rectangular) point 12
11. Lip terminally with a broad and conspicuous rectangular,
    rhomboid or obtriangular, often dentate appendage 19

12. Lip markedly saddle-shaped. Column tapering towards the
    base (in side view); stigmatic cavity distinctly longer than wide
    . . . . 15. **O. bertolonii**
12. Lip straight to moderately saddle-shaped. Column not
    tapering towards the base (in side view); stigmatic cavity
    approximately as long as wide 13

13. Lip moderately saddle-shaped; its terminal point erect. . . . .
    . . . . H5. **O. ×flavicans**
13. Lip straight; its terminal point porrect or pointing downwards,
    occasionally absent 14

14. Petals linear-lanceolate. Lip deeply three-lobed; side lobes
    and the sides of the mid-lobe reflexed and thus making the
    lip appear entire and narrowly oblong when seen from above;
    mid-lobe with a broad yellow to light brown margin; mirror
    bluish grey to reddish brown, never framed with a white
    border, in most cases completely isolated from the base of
    the lip . . . . 17. **O. lunulata**

14. Without the above combination of character states 15

15. Mirror connected with the base of the lip by distinct broad bands (or mirror absent) 16

15. Mirror isolated from the base of the lip or only connected with the base by delicate lines 17

16. Lip entire to moderately (rarely deeply) three-lobed near the middle; bulges, if present, distinctly isolated from the margin of the lip (side lobes flat, if recognisable)
. . . . 16. **O. sphegodes** [but see note 3]

16. Lip moderately to deeply three-lobed near the base; side lobes converted into hump-shaped to obliquely conical bulges . . . .
. . . . 19. **O. kotschyi**

17. Lip deeply three-lobed; side lobes converted into (often weakly developed) bulges; mirror often white . . . . . . 18. **O. reinholdii**

17. Lip entire to moderately three-lobed; bulges (if present) distinctly isolated from the margin of the lip; mirror never white 18

18. Petals nearly glabrous. Lip with (purplish) black to dark purplish brown ground colour, dark velvety along the margin
. . . . 14. **O. ferrum-equinum**

18. Petals shaggy to velvety. Lip with (reddish) brown to yellowish brown or olive-green ground colour, in its basal part more or less shaggy of usually light hairs along the margin . . . . 13. **O. argolica** [but see note 4]

19. Lip entire (to shallowly three-lobed); bulges weakly developed to obliquely conical, distinctly isolated from the margin of the lip . . . . 12. **O. fuciflora** [but see note 5]

19. Lip deeply three-lobed; side lobes converted into obliquely conical to horn-shaped (in rare cases only weakly developed) bulges 20

20. Side lobes of the lip densely and whitish shaggy on their outer side; mirror often without connection to the base of the lip . . . . H3. **O. ×delphinensis**

20. Side lobes of the lip brownish shaggy on their outer side; mirror always with connection to the base of the lip 21

21. Dorsal sepal flat to shallowly boat-shaped, more or less incurved, from the base describing an obtuse angle to the column (and consequently not forming a roof over the latter). Petals shaggy, spreading to slightly recurved . . . .
. . . . 11. **O. scolopax** [but see note 6]

21. Dorsal sepal more or less boat-shaped, strongly incurved, from the base nearly parallel to the column (and consequently forming a roof over the latter). Petals velvety, recurved .... 10. **O. umbilicata**

NOTES

1. If the plant does not match any subspecies of O. *fusca*, see H1. O. ×*brigittae*.

2. If the plant does not match any subspecies of O. *omegaifera*, see H1. O. ×*brigittae*.

3. If the plant does not match any subspecies of O. *sphegodes*, see H4. O. ×*arachnitiformis* or H5. O. ×*flavicans*.

4. If the plant does not match any subspecies of O. *argolica*, see H3. ×O. *delphinensis*.

5. If the plant does not match any subspecies of O. *fuciflora*, see I12. O. ×*vicina* or H4. O. ×*arachnitiformis*.

6. If the plant does not match any subspecies of O. *scolopax*, see H2. O. ×*vicina*.

1. *Ophrys atlantica*

Ophrys atlantica can be regarded as a tourist on the Spanish *Costa del Sol*; only in southernmost Andalusia does this North African orchid look into Europe. The species is mainly characterised by the wavy petals and the saddle-shaped lip with a slender, narrow base. It varies very little and can only be confused with certain forms of *O. fusca* (particularly subsp. *iricolor*). Both in *O. atlantica* and *O. fusca* subsp. *iricolor* the lip is wine-red underneath, but in the latter the lip is neither saddle-shaped nor with a slender, narrow base.

Ophrys atlantica is pollinated by the bee *Chalicodoma parietina* (family Megachilidae). It shares this pollinator with the central Mediterranean *O. bertolonii*, and the saddle-shaped lip in both species probably represents a common, independent adaptation to this particular bee.

FIG. 27A

FIG. 27. *Ophrys atlantica* (**A-C.** Spain, Andalusia, Casa el Cura, 15th April 2001). Photos by H. Æ. Pedersen.

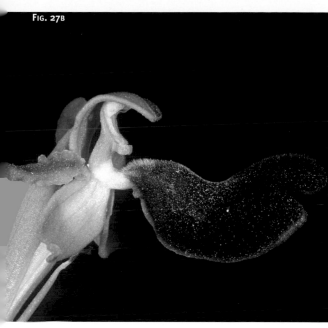

FIG. 27B

FIG. 27C

HABITAT. In full sunlight to light shade on basic, dry to somewhat moist ground from sea level to 1500 m altitude. Typical habitats include open deciduous woods and pine woods, garrigue and scrubby grasslands.

FLOWERING TIME. From March to June – early in Andalusia and late in the mountains of North Africa.

DISTRIBUTION. Western Mediterranean; almost limited to the northern parts of Morocco and Algeria (and possibly Tunisia where the hybrid *O. atlantica × fusca* subsp. *fusca* has been photographed). In Andalusia, it occurs very locally in the province of Malaga. [Spa; Mor] MAP 3.

DESCRIPTION. *Ophrys atlantica* Munby
Syn. *O. fusca* Link subsp. *atlantica* (Munby) E. G. Camus; *O. fusca* Link subsp. *durieui* (Rchb.f.) Soó.

Plant slender, (10-)15-30(-35) cm tall with 1-4 flowers in a lax spike. **Sepals** pale green to yellowish green, (narrowly) ovate to elliptic, 11-18 × 5-8.5 mm; dorsal sepal flat to shallowly boat-shaped, incurved, from the base nearly parallel to the column. **Petals** olive-green to olive-brown, linear-oblong with wavy margins, 8.5-15 × 2-4 mm, glabrous, recurved. **Lip** with purplish-black ground colour, though somewhat paler on the slender, stalk-like base, saddle-shaped with recurved sides, three-lobed near the apex, 15-22(-25) × 12-22 mm, velvety except on the mirror; bulges absent; mid-lobe indistinctly longer than the side lobes, emarginate; **mirror**

67

Map 3. The European range of *Ophrys atlantica*. This species is not subdivided into subspecies.

distinct, consisting of an irregularly rectangular to elliptic figure on the subbasal part of the lip (but with no connection to the absolute base of the lip), shining blue. **Column** rounded, not tapering towards the base (in side view); stigmatic cavity at least as long as wide and approximately twice as wide as the anther, devoid of lateral, eye-like knobs at base. Fig. 27.

References: Baumann (1975), Baumann & Künkele (1982, 1988), Buttler (1986), Davies et al. (1983), Delforge (1994, 2001, 2005), Nelson (1962), Paulus & Gack (1981, 1983), Sundermann (1980).

2. *Ophrys omegaifera*

"Omegaifera" indicates that this species carries the Greek letter omega (ω). This alludes to the circumstance that the mirror of the lip is delimited by an ω-shaped band. The lip in *O. omegaifera* generally resembles a tiny boxing glove, but otherwise the flowers vary substantially. The species might be confused with *O. fusca* which now and then carries a (less conspicuous) ω-

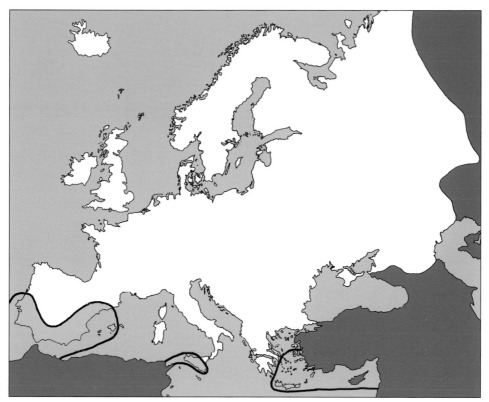

Map 4. The disjunct distribution of *Ophrys omegaifera* in Europe. Five subspecies are recognized; subsp. *dyris* and subsp. *hayekii* occupy the westernmost and the central range, respectively, whereas subsp. *omegaifera*, subsp. *fleischmannii* and subsp. *israelitica* are restricted to the easternmost range.

shaped figure. The latter species, however, always has a distinct longitudinal furrow at the base of the lip, a feature that is never found in O. *omegaifera*. Nevertheless, identification problems may be encountered at both the eastern and the western end of the Mediterranean, as O. *omegaifera* is in both of these areas involved in a partly stabilised hybrid complex with O. *fusca* (see H1. O. ×*brigittae*).

The morphological variation mainly falls between five groups of populations that are also mutually separated by widely different geographic ranges and/or different specific pollinators of the bee genus *Anthophora* (family Anthophoridae). Based on the patterns of variation in O. *omegaifera* we recognise five subspecies.

Habitat. In full sunlight or light shade on dry to moist, basic to slightly acid ground from sea level to 2000 m altitude. Typical habitats include garrigue, grassland, open pine and cypress woods, pesticide-free olive groves and fallow fields.

FLOWERING TIME. From January to early May, being at its peak from mid–March to mid–April.

DISTRIBUTION. In the eastern and the westernmost parts of the Mediterranean as well as in a small area encompassing Sicily and northeastern Tunisia. [Aeg, Bal, Cre, Gre, Por, Sic, Spa; Ana, Cyp, Isr, Mor, Tun] MAP 4.

DESCRIPTION. *Ophrys omegaifera* H. Fleischm.
Synonyms are listed under the subspecies.

Plant compact to slender, 8-35(-50) cm tall with 1-8(-10) flowers in a relatively dense to lax spike. **Sepals** (pale) green to yellowish green, ovate to elliptic, 9-20 × 6-11 mm; dorsal sepal boat-shaped, incurved, from the base nearly parallel to the column. **Petals** (yellowish) green, often suffused with red or brown (especially along the margins), narrowly oblong with (almost) flat margins, (4-)7-14 × 2-5 mm, glabrous, spreading to moderately incurved. **Lip** with brown to grey or almost black ground colour, straight to abruptly downcurved with recurved margin, three-lobed around the middle or closer to the apex, (9-)10-28.5 × 8-27 mm, velvety to shaggy (especially towards the margin); bulges absent; mid-lobe distinctly longer than the side lobes, emarginate; **mirror** fairly distinct, consisting of a figure shaped more or less like the tail of a fish, stretching from the base to over the middle of the lip, dull brown to violet or bluish grey (sometimes marbled), in front delimited by a conspicuous (in subsp. *hayekii* often obscure) white to blue ω-shaped band and sometimes with a pale area at the base. **Column** rounded, not tapering towards the base (in side view); stigmatic cavity usually as wide as long and approximately twice as wide as the anther, devoid of lateral eye-like knobs at base. FIGS 28-32.

Key to the subspecies

1. Lip straight (to slightly and gradually downcurved) at base; the ω-shaped band along the front edge of the mirror distinct or obscure 2
1. Lip abruptly downcurved at base; the ω-shaped band along the front edge of the mirror distinct 3

2. Lateral sepals less than 1.75 times as long as wide; broadest at or above the middle. Mirror usually marbled . . . . 2.5. **subsp. *israelitica***
2. Lateral sepals more than 1.75 times as long as wide; broadest at or below the middle. Mirror not marbled . . . . 2.3. **subsp. *hayekii***

3. Mid-lobe and side lobes of the lip short-hairy; the ω-shaped band of the mirror (greyish) blue in newly opened flowers . . . . .2.1. **subsp.** *omegaifera*
3. Mid-lobe and side lobes of the lip long-hairy; the ω-shaped band of the mirror white to light grey (rarely light bluish) in newly opened flowers 4

4. Flowers inclined-spreading. Mirror reddish-brown (to cream or greenish), pale at base . . . . 2.2. **subsp.** *dyris*
4. Flowers erect. Mirror usually bluish, not pale at base . . . . 2.4. **subsp.** *fleischmannii*

2.1. *Ophrys omegaifera* subsp. *omegaifera*

This subspecies is particularly frequent in Crete and Rhodes. In Crete it is so variable in flower size and flowering time that some authors treat it as two separate species. Early-flowering plants with large, often more bluish grey lips have thus been described as *O. basilissa*. However, we only recognise this entity at varietal level. The flowering times overlap, and where the two varieties grow in mixed colonies they often form hybrid swarms that render identification difficult. Flowering takes place from January to April.

The distribution primarily encompasses Crete (particularly the central and eastern parts), Karpathos (southern part), and Rhodes (where subsp. *omegaifera* is fairly common). Additional occurrences are known from southwestern Anatolia and from the following Aegean islands: Andros, Arkoi, Chios, Kalimnos, Kea, Kos, Leros, Lipsoi, Naxos, Paros, Patmos, Samos and Skiros [Aeg, Cre, Gre; Ana].

DESCRIPTION. *Ophrys omegaifera* H. Fleischm. subsp. *omegaifera*
Syn. [var. *omegaifera*]: *O. fusca* Link subsp. *omegaifera* (H. Fleischm.) E. Nelson
Syn. [var. *basilissa* (Alibertis & H. R. Reinhard) N. Faurholdt]: *O. basilissa* Alibertis & H. R. Reinhard; *O. omegaifera* H. Fleischm. subsp. *basilissa* (Alibertis & H. R. Reinhard) H. Kretzschmar.

Plant compact (to slender), 8-25(-30) cm tall with 1-4(-8) inclined-spreading to erect flowers in a relatively dense (very rarely lax) spike. **Sepals** 11-16 × 6-9.5 mm; the lateral ones 1.5-2.5 times as long as wide, broadest below, at, or above the middle. **Petals** (5-)7-14 × 2-5 mm. **Lip** with brown to (purplish) grey (rarely bluish) ground colour, opposite the hind edge of the side lobes abruptly downcurved, (13-)15-28.5 × 14-27 mm; mid-lobe and side lobes short-hairy; **mirror** usually not marbled, orange-brown to

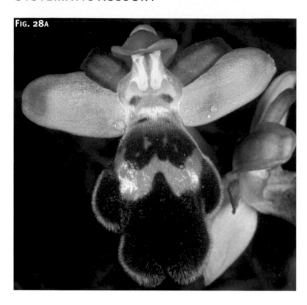

FIG. 28A

FIG. 28. *Ophrys omegaifera* subsp. *omegaifera*. **A-C.** var. *omegaifera*; **D.** var. *basilissa* (**A.** Greece, Samos, Agia Kyriaki, 11th April 1999; **B.** Greece, Crete, Rodovani, 14th April 2003; **C.** Greece, Naxos, Koronos, 16th April 2000; **D.** Greece, Crete, Sougia, 23rd March 1997). Photos by H. Æ. Pedersen (**A-B**), A. Gøthgen (**C**), N. Faurholt (**D**).

blue or bluish violet, not pale at the base, the ω-shaped band in front sharply delimited and (in newly opened flowers) (greyish) blue. FIG. 28.

Var. *omegaifera* is mainly characterised by its 13-21 mm long lip with light brown (rarely light grey or bluish) ground colour. **Plant** 8-20(-25) cm tall with 1-4(-8) flowers in a dense (very rarely lax) spike. **Sepals** 12-17 × 6-9.5 mm. **Petals** (5-)7-11 × 2-4 mm. **Lip** with light brown (rarely light grey or bluish ground colour), (13-)15-21 × 14-23 mm; **mirror** (orange-)brown (rarely bluish or whitish). [FIG. 28A-C]. This variety is relatively late-flowering (February-April). Var. *omegaifera* occurs throughout the geographic range of the subspecies. [Aeg, Cre, Gre; Ana].

FIG. 28C

FIG. 28D

Var. *basilissa* is mainly characterised by its 17.5-28.5 mm long lip with dark (purplish) grey (less frequently dark brown) ground colour. **Plant** 10-25(-30) cm tall with 1-4(-6) flowers in a relatively dense spike. **Sepals** 13-20 × 7-11 mm. **Petals** 10-14 × 3-5 mm. **Lip** with dark (purplish) grey (less frequently dark brown) ground colour, 17.5-28.5 × 18-27 mm; **mirror** blue to bluish-violet (less frequently brown or whitish). [FIG. 28D]. This variety flowers very early (January-March). Var. *basilissa* is endemic to Greece, where it is known from central and eastern Crete as well as from Kalimnos, Leros, and Paros [Aeg, Cre, Gre].

2.2. *Ophrys omegaifera* subsp. *dyris*

This subspecies is distinguished by its flowers, which are often more rich in contrast than the flowers of the other subspecies. Especially in the Portuguese province of Algarve plants can be encountered that have generally more colourful flowers and an almost straight lip. Subsp. *dyris* is the only representative of *O. omegaifera* in the western Mediterranean and can only be confused with hybrids involving itself. Not least on Mallorca and in mainland Spain hybridisation and backcrossing with *O. fusca* result in bewildering intermediary forms (see H1. *O. ×brigittae*). Peak flowering is in March-April. In Europe, flowering specimens are now and then encountered much earlier, whereas in the mountains of North Africa, the flowering is prolonged into May.

The distribution is western Mediterranean and is in Europe limited to Mallorca and the southwestern part of the Iberian peninsula (Estremadura, Ribatejo, Algarve and Andalusia). [Bal, Por, Spa; Mor].

DESCRIPTION. *Ophrys omegaifera* H. Fleischm. subsp. *dyris* (Maire) Del Prete Syn. *O. algarvensis* D. Tyteca et al.; *O. dyris* Maire.

Plant slender, 10-35(-50) cm tall with (1-)3-8 inclined-spreading flowers in a relatively lax (to dense) spike. **Sepals** 10-17 × 6-8 mm; the lateral ones 1.5-2.5 times as long as wide, broadest below, at or above the middle. **Petals** 7-12 × 2-3.5 mm. **Lip** with dark grey to blackish brown ground colour, opposite the hind edge of the side lobes abruptly downcurved, 10-19 × 9-14 mm; mid-lobe and side lobes long-hairy; **mirror** usually not marbled, reddish brown (to cream or greenish), pale at the base, the ω-shaped band in front sharply delimited and (in newly opened flowers) white to light grey (rarely light bluish). FIG. 29.

FIG. 29. *Ophrys omegaifera* subsp. *dyris* (**A-B.** Spain, Andalusia, Grazalema, 18th April 2001; **C.** Portugal, the Algarve, Malhao, 4th April 1999). Photos by H. Æ. Pedersen (**A-B**), N. Faurholdt (**C**).

75

2.3 *Ophrys omegaifera* subsp. *hayekii*

Subsp. *hayekii* differs from all other subspecies of *O. omegaifera* in its usually obscure ω–shaped band along the front edge of the mirror and in its (very shallow!) longitudinal furrow at the base of the lip. In our opinion, however, these deviant features are so insignificant that the plant can be assigned to *O. omegaifera* without reservation. It grows on calcareous ground in grassland, garrigue and light-open edges of woods, up to 900 m altitude. This subspecies is fairly late-flowering, from mid-April to early May.

Subsp. *hayekii* is extant, though very rare, in central and southeastern Sicily. Former occurrences at Palermo and in northeastern Tunisia appear to be extirpated. [Sic; Tun].

DESCRIPTION. *Ophrys omegaifera* H. Fleischm. subsp. *hayekii* (H. Fleischm. & Soó) Kreutz
Syn. *O. mirabilis* Geniez & Melki.

Plant compact to slender, 15-22 cm tall with 2-5(-6) inclined-spreading flowers in a relatively lax spike. **Sepals** 10-15 × 5-7 mm; the lateral ones 1.8-2.8 times as long as wide, broadest at or below the middle. **Petals** 7-10 × 1.5-2.5 mm. **Lip** with reddish brown to blackish brown ground colour,

FIG. 30. *Ophrys omegaifera* subsp. *hayekii* (**A-B.** Italy, Sicily, Siracusa, 27th April 2004). Photos by C. A. J. Kreutz.

FIG. 30B

opposite the hind edge of the side lobes straight (to slightly and gradually downcurved), 12-18 × 8-10 mm; mid-lobe and side lobes short-hairy; **mirror** not marbled, bluish grey, sometimes pale at the base, the ω-shaped band in front (if present) not sharply delimited, (in newly opened flowers) light bluish grey to whitish. FIG. 30.

2.4. *Ophrys omegaifera* subsp. *fleischmannii*

In most areas this subspecies is easily recognised. In Crete, however, mixed colonies of subsp. *fleischmannii* and subsp. *omegaifera* cause confusion, and the picture may also be blurred by the occurrence of hybrids. Flowering takes place from February (occasionally January) to April.

Subsp. *fleischmannii* is endemic to Greece, where it is known from Crete and a few Cycladean islands. Finds reported from Idhra and the province of Attika on the mainland have not been documented. [Cre, Gre].

DESCRIPTION. *Ophrys omegaifera* H. Fleischm. subsp. *fleischmannii* (Hayek) Del Prete.
Syn. *O. fleischmannii* Hayek.

Plant compact, 8-15(-20) cm tall with 2-5(-8) erect flowers in a relatively dense spike. **Sepals** 11-15 × 6-9.5 mm; the lateral ones 1.5-2.5 times as long as wide, broadest below, at or above the middle. **Petals** 8-11 × 2-3.5 mm. **Lip** with blackish brown to dark grey ground colour, opposite the hind edge of the side lobes abruptly downcurved, 13-17 × 10-16 mm; mid-lobe and side lobes long-hairy; **mirror** usually not marbled, bluish violet (to brown), not pale at the base, the ω-shaped band in front sharply delimited and (in newly opened flowers) white to light grey (rarely light bluish). FIG. 31.

2.5. *Ophrys omegaifera* subsp. *israelitica*

Subsp. *israelitica* is not difficult to tell apart from other subspecies of *O. omegaifera*. It is therefore surprising that it was not described until 1988 (as a separate species). Prior to this, populations of this entity were referred to subsp. *fleischmannii*, a circumstance that should be remembered when consulting the literature published before 1988. Now and then subsp. *israelitica* grows in slightly acid soil. The flowering time is very early, often starting around Christmas in Israel. The peak is in February-March, and on Naxos flowering specimens can still be found in April.

The distribution encompasses the easternmost part of the Mediterranean, particularly the Levant and Cyprus. In Europe subsp. *israelitica* is very rare

FIG. 31. *Ophrys omegaifera* subsp. *fleischmanii* (Greece, Crete, Rodovani, 14th April 2003).
Photo by H. Æ. Pedersen.

FIG. 32. *Ophrys omegaifera* subsp. *israelitica* (**A.** Greece, Naxos, Kato Sagri, 16th April 2000; **B.** Cyprus, southern part, Pegeia, 17th March 2001). Photos by A. Gøthgen (**A**), N. Faurholdt (**B**).

and only reported from the following Cycladean islands: Naxos, Paros and Siros. [Gre; Ana, Cyp, Isr].

DESCRIPTION. *Ophrys omegaifera* H. Fleischm. subsp. *israelitica* (H. Baumann & Künkele) G. & K. Morschek
Syn. *O. israelitica* H. Baumann & Künkele.

Plant slender, (8-)12-20(-35) cm tall with 2-8(-10) inclined-spreading flowers in a relatively lax (to fairly dense) spike. **Sepals** 9-13 × 6-8.5 mm; the lateral ones 1.5-1.7 times as long as wide, broadest at or above the middle. **Petals** (4-)7-9.5 × 2-3.5 mm. **Lip** with brown to grey ground colour, opposite the hind edge of the side lobes straight (to slightly and gradually downcurved), (9-)11-16 × 8-13 mm; mid-lobe and side lobes short-hairy; **mirror** in general distinctly marbled, pale bluish grey (to brownish), not pale at the base, the ω-shaped band in front sharply or obscurely delimited, (in newly opened flowers) greyish blue. FIG. 32.

REFERENCES: Alibertis et al. (1990), Baumann & Dafni (1981), Baumann & Künkele (1982, 1988), Buttler (1986), E. Danesch & O. Danesch (1969), Davies et al. (1983), Delforge (1994, 2001, 2005), Nelson (1962), Paulus (1988), Paulus & Gack (1981, 1983, 1986, 1990a, 1992), Sundermann (1980).

3. *Ophrys fusca*

Ophrys fusca is one of the commonest bee orchids in the Mediterranean and one of the most variable and discussed species of the genus. In recent years, studies of pollinator specificity, in particular, have lead to attempts at a very fine splitting of *O. fusca* at species level. However, the vast majority of these "species" are poorly distinguished morphologically (FIG. 26) and do not, in our opinion, deserve formal systematic recognition. Both the alleged pollinator specificity and morphological differences are in need of further documentation. However, our present classification may prove to be slightly too conservative on this point; we suspect that modest splitting of the present subsp. *fusca* might be justified when more rigorous data have been obtained.

Ophrys fusca is reminiscent of *O. lutea* and *O. omegaifera*. It differs from *O. lutea* in the recurved margins of the lip and from *O. omegaifera* in the distinct longitudinal furrow basally on the lip. Identification can be difficult in the western and eastern parts of the Mediterranean where *O. fusca* is involved in a partly stabilised hybrid complex with *O. omegaifera* (see H1. *O. ×brigittae*).

Within *O. fusca*, as circumscribed here, the morphological variation mainly falls between five groups of populations, which we recognise as distinct subspecies. All of these have well-defined geographic ranges and are adapted to different pollinators. Subsp. *blitopertha* is unique among bee orchids, as it is the only *Ophrys* that is consistently pollinated by a beetle (i.e. the chafer *Blitopertha lineolata*, family Scarabaeidae). Subsp. *cinereophila* and subsp. *pallida* seem to be specifically adapted to pollination by two andrenid bees (i.e. *Andrena cinereophila* and *A. orbitalis*, respectively). Both subsp. *fusca* and subsp. *iricolor* are less specific, but their respective selections of pollinators appear not to overlap; nor do they include any of the above pollinators.

HABITAT. Dry to moist, occasionally wet, soil in full sunlight to light shade from sea level to 1500 m altitude. Typical habitats include roadside verges, garrigue, maquis, pesticide-free olive-groves and grassland as well as light-open pine and oak woods.

FLOWERING TIME. From January to June, but peaks in March-April. Now and then, flowering individuals can be encountered already in December.

DISTRIBUTION. Throughout the Mediterranean and along the French Atlantic coast north to Brittany. [Aeg, Alb, Bal, Cor, Cre, Fra, Gre, Ita, Mal, Por, Sar, Sic, Spa, Tur, Yug; Ana, Cyp, Isr, Mor, Tun] MAP 5.

DESCRIPTION. *Ophrys fusca* Link.

Synonyms are listed under the subspecies.

Plant compact to slender, (5-)7-40(-70) cm tall with 1-10(-11) flowers in a lax to dense spike. **Sepals** (pale) green to yellowish green, olive-green or white, ovate to elliptic, 7-18 × 3-10 mm; dorsal sepal more or less boat-shaped, incurved, from the base nearly parallel to the column. **Petals** (yellowish) green (to yellow), often suffused with red or brown, narrowly oblong to linear-oblong with (almost) flat margins, 4-12 × 1-4 mm, glabrous, spreading to moderately incurved. **Lip** with (purplish) brown (to purplish black) ground colour and sometimes a yellow or yellowish green margin, straight (to gradually downcurved) or with a knee-like bend at the base and always with recurved margins, distinctly longitudinally furrowed at base, three-lobed around the middle or closer to the apex, 7-26 × 5-21 mm, velvety (especially towards the margin); bulges absent; mid-lobe distinctly longer than the side lobes, emarginate (to rounded); **mirror** distinct to somewhat obscure, consisting of a figure shaped more or less like the tail of a fish (sometimes longitudinally split) and extending from the base of the lip to above the middle, dull grey to (occasionally shining) blue or

Map 5. The European range of *Ophrys fusca*. Five subspecies are recognized in Europe, where the highest diversity is found in Greece (four subspecies).

dull violet to reddish brown (rarely marbled), often more strongly coloured in front. **Column** rounded, not tapering towards the base (in side view); stigmatic cavity wider than long and approximately twice as wide as the anther, devoid of lateral eye-like knobs at base. FIGS 33-37.

Key to the subspecies

1. Lip wine-red underneath; mirror shining, sharply delimited
. . . . 3.2. **subsp.** *iricolor*

1. Lip (pale) green underneath; mirror dull, not very sharply delimited 2

2. Lip straight (to more or less downcurved for most of its length) 3
2. Lip with a moderate to strong knee-like bend close to its base 4

3. Lip slightly to strongly downcurved (rarely straight), in cross section distinctly vaulted, with or without a narrow (rarely broad) yellow margin; side lobes more or less downcurved . . . . 3.1. **subsp.** *fusca*
3. Lip straight, in cross section nearly flat, with an often broad yellow (to yellowish brown) margin; side lobes spreading . . . .
. . . . 3.4. **subsp.** *blitopertha*

4. Flowers spirally arranged on the uppermost one third (or more) of the stem. Sepals (pale) green . . . . 3.5. **subsp.** *cinereophila*
4. Flowers densely clustered on the uppermost one fourth (or less) of the stem, not forming a distinct spiral. Sepals white (occasionally with a greenish, yellowish or rose-coloured tinge) . . . .
. . . . 3.3. **subsp.** *pallida*

3.1. *Ophrys fusca* subsp. *fusca*

Employing our wide circumscription of *O. fusca* subsp. *fusca*, this subspecies is extremely variable, and certain forms may be confused with other subspecies. For instance, this is true for small-flowered plants that resemble subsp. *cinereophila* (in which, however, the flowers are spirally arranged in a denser inflorescence) and large-flowered plants that resemble subsp. *iricolor* (in which, however, the lip is wine-red underneath). Flowering takes place from January to June, in most areas with a peak in March-April. In coastal areas, flowering individuals can sometimes be encountered in late December.

Subsp. *fusca* is distributed throughout the range of the species, except in the easternmost areas. [Aeg, Alb, Bal, Cor, Cre, Fra, Gre, Ita, Mal, Por, Sar, Sic, Spa, Tur, Yug; Ana, Mor, Tun].

Fig. 33A

Fig. 33B

Fig. 33C

DESCRIPTION. *Ophrys fusca* Link subsp. *fusca*
Syn. *O. arnoldii* P. Delforge; *O. attaviria* D. Rückbrodt & Wenker; *O. bilunulata* Risso; *O. caesiella* P. Delforge; *O. calocaerina* J. Devillers-Terschuren & P. Devillers; *O. creberrima* Paulus; *O. fusca* Link subsp. *creberrima* (Paulus) H. Kretzschmar; *O. cressa* Paulus; *O. fusca* Link subsp. *cressa* (Paulus) H. Kretzschmar; *O. creticola* Paulus; *O. fusca* Link subsp. *creticola* (Paulus) H. Kretzschmar; *O. eptapigiensis* Paulus; *O. fabrella* Paulus & Ayasse ex P. Delforge; *O. flammeola* P. Delforge; *O. forestieri* (Rchb.f.) Lojac.; *O. funerea* Viv.; *O. fusca* Link subsp. *funerea* (Viv.) Arcang. [non (Viv.) Nyman, comb. superfl.]; *O. gackiae* P. Delforge; *O. hespera* J. Devillers-Terschuren & P. Devillers; *O. laurensis* Geniez & Melki; *O.*

Fig. 33. *Ophrys fusca* subsp. *fusca* (**A.** Italy, Sicily, Buccheri, 17th April 1999; **B-C.** Greece, Rhodes, Profitis Ilias, 15th April 1996; **D.** Greece, Crete, Melambes, 25th March 1997; **E.** Greece, Chios, Kato Fano, 17th April 2005). Photos by H. Æ. Pedersen (**A-B, E**), N. Faurholdt (**C-D**).

leucadia Renz; *O. fusca* Link subsp. *leucadia* (Renz) H. Kretzschmar; *O. lindia* Paulus; *O. lojaconoi* P. Delforge; *O. lucana* P. Delforge et al.; *O. lucentina* P. Delforge; *O. lucifera* J. Devillers–Terschuren & P. Devillers; *O. lupercalis* J. Devillers–Terschuren & P. Devillers; *O. marmorata* G. & W. Foelsche; *O. mesaritica* Paulus et al.; *O. obaesa* Lojac.; *O. ortuabis* M. P. Grasso & L. Manca; *O. parosica* P. Delforge; *O. parvula* Paulus; *O. pectus* Mutel; *O. peraiolae* G. Foelsche et al.; *O. perpusilla* J. Devillers–Terschuren & P. Devillers; ?*O. persephonae* Paulus; *O. punctulata* Renz; *O. sabulosa* Paulus & Gack ex P. Delforge; *O. sulcata* J. Devillers–Terschuren & P. Devillers; *O. thriptiensis* Paulus; *O. fusca* Link subsp. *thriptiensis* (Paulus) H. Kretzschmar; *O. zonata* J. Devillers–Terschuren & P. Devillers.

85

Plant compact to slender, (5-)10-35(-40) cm tall with 2-10(-11) alternate flowers in a lax (to fairly dense) spike on the uppermost one third (or more) of the stem. **Sepals** pale green to yellowish green, 7-16 × 3-9 mm. **Petals** olive-green to yellowish green, often suffused with red or brown, 4-12 × 1-4 mm. **Lip** (pale) green underneath, slightly to strongly downcurved for most of its length (rarely straight), in cross section vaulted, 7-23 × 5.5-21 mm, brown right up to the margin or with a very narrow to relatively broad yellow to yellowish green margin; side lobes more or less downcurved; **mirror** often somewhat obscure, dull bluish violet to greyish (rarely reddish or marbled). FIG. 33.

3.2. *Ophrys fusca* subsp. *iricolor*

Subsp. *iricolor* is easily recognised (see, however, remarks under subsp. *fusca*) and exhibits only little variation. In the western Mediterranean, the plants are in general slightly taller and often provided with a fine yellowish green margin on the lip. Flowering takes place from February to May, and usually peaks in April. The first populations to flower are those in Israel and Cyprus, where subsp. *iricolor* has often stopped flowering by late March.

This subspecies mainly occurs in the eastern Mediterranean, but it is also well represented in Tunisia and Sardinia. In fact, the total distribution

FIG. 34A

FIG. 34. *Ophrys fusca* subsp. *iricolor* (**A.** Greece, Samos, Agia Kyriaki, 11th April 1999; **B.** Greece, Samos, Agia Triada, 12th April 1999; **C.** Tunisia, Djebel Zaghouan, 15th March 1998). Photos by H. Æ. Pedersen.

seems to be wider than usually acknowledged. For instance, we found subsp. *iricolor* in Portugal (Algarve) in 1999, and records from Malta and Mallorca appear reliable. Finds reported from mainland Italy (Monte Argentario), Sicily and mainland Spain (Andalusia) should be checked up on. [Aeg, Bal, Cor, Cre, Gre, Isr, Mal, Por, Sar; Ana, Cyp, Tun].

DESCRIPTION. *Ophrys fusca* Link subsp. *iricolor* (Desf.) K. Richt. Syn. ?*O. astypalaeica* P. Delforge; *O. eleonorae* J. Devillers-Terschuren & P. Devillers; *O. iricolor* Desf. subsp. *eleonorae* (J. & P. Devillers-Terschuren & P. Devillers) Paulus & Gack; *O. iricolor* Desf. s.s.

Plant slender, 10-40(-70) cm tall with 1-7 alternate flowers in a relatively lax spike on the uppermost one third (or more) of the stem. **Sepals** (pale) green, 10-18 × 6-10 mm. **Petals** (olive-)green to purplish brown, sometimes suffused with ochre-yellow, 6-12 × 2-4 mm. **Lip** wine-red underneath, straight, in cross section vaulted, 14-26 × 10-21 mm, brown right up to the margin or with a very narrow yellowish green margin; side lobes more or less downcurved; **mirror** sharply delimited, shining blue (occasionally a bit marbled), normally turning into purple in its bipartite base. FIG. 34.

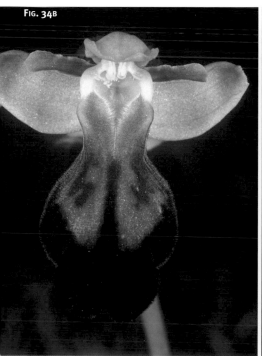

FIG. 34B

FIG. 34C

3.3. *Ophrys fusca* subsp. *pallida*

This subspecies is very characteristic and easily distinguished from the others. Flowering takes place from late February to early May, with a peak from mid-March to mid-April.

Subsp. *pallida* occupies a very narrow range. Most well-known are the occurrences in the neighbourhood of Palermo in northwestern Sicily, but additional populations are found in northeastern Algeria. Records from Tunisia, Malta and Sardinia should be checked up on. [Sic].

DESCRIPTION. *Ophrys fusca* Link subsp. *pallida* (Raf.) E. G. Camus et al. Syn. *O. pallida* Raf.

Plant slender, 10-30 cm tall with 2-6 alternate flowers in a dense spike on the uppermost one fourth (or less) of the stem. **Sepals** white (occasionally with a greenish, yellowish or rose-coloured tinge), 7-10 × 4.5-6.5 mm. **Petals** yellow to olive-green, 5-7 × 2-3 mm. **Lip** (pale) green underneath, with a strong knee-like bend at the base, in cross section vaulted, 7-11 × 5-9.5 mm, brown right up to the margin or with a very narrow yellowish margin; side lobes strongly recurved, almost hidden below the lip; **mirror** somewhat obscure, dull bluish to greyish. FIG. 35.

FIG. 35. *Ophrys fusca* subsp. *pallida* (**A-B.** Italy, Sicily, Bosco di Ficuzza, 20th April 1999). Photos by H. Æ. Pedersen.

Fig. 36A

Fig. 36B

Fig. 36. *Ophrys fusca* subsp. *blitopertha* (**A**. Greece, Chios, Kato Fano, 17th April 2005; **B**. Greece, Samos, Ormos Marathokampos, 8th April 1998). Photos by H. Æ. Pedersen (**A**), N. Faurholdt (**B**).

3.4. *Ophrys fusca* subsp. *blitopertha*

This subspecies may be confused with subsp. *fusca*, but it differs in its straight, nearly flat and usually horizontally orientated lip with large spreading side lobes, broad mid-lobe and (usually broad) yellow margin. According to our observations from Chios, Lesbos, Samos, Rhodes and Naxos, subsp. *blitopertha* generally prefers garrigue and old, pesticide-free olive-groves. Flowering is relatively late and takes place from late March to early May.

This subspecies is distributed in areas of the eastern Mediterranean and is particularly frequent on some of the Aegean islands (so far reported from Chios, Fournoi, Kalimnos, Lesbos, Naxos, Patmos, Rhodes and Samos). [Aeg, Gre; Ana].

DESCRIPTION. *Ophrys fusca* Link subsp. *blitopertha* (Paulus & Gack) N. Faurholdt & H. A. Pedersen
Syn. *O. blitopertha* Paulus & Gack.

Plant compact, 7-20 cm tall with 1-4(-7) alternate flowers in a relatively dense spike on the uppermost one third (or more) of the stem. **Sepals** pale green to olive-green, 8-13 × 4-7 mm. **Petals** (olive-)green, often suffused with brown, 5-10 × 2-3 mm. **Lip** (pale) green underneath, straight, in cross section nearly flat, 8-17 × 7.5-15 mm, with a usually broad yellow (to yellowish brown) margin; side lobes spreading; **mirror** somewhat obscure, dull and dark reddish brown to bluish grey. FIG. 36.

FIG. 37. *Ophrys fusca* subsp. *cinereophila* (**A.** Greece, Crete, Nea Kria Vrisi, 13th April 2003; **B.** Greece, Chios, Avgonima, 14th April 2005). Photos by H. Æ. Pedersen.

3.5. *Ophrys fusca* subsp. *cinereophila*

Subsp. *cinereophila* may be confused with small-flowered forms of subsp. *fusca*, but it differs in the spirally arranged flowers, the lip of which is often provided with a distinct yellow margin. Flowering takes place from late February to mid–April, usually being at its best in the second half of March.

The distribution is eastern Mediterranean, but insufficiently known, as this plant was not recognised systematically until 1998. Reliable records exist from parts of mainland Greece, a number of Aegean islands, Cyprus, Syria and areas in western Anatolia. Personally, we have seen subsp. *cinereophila* in Chios, Crete, Rhodes, Samos and Cyprus. [Aeg, Cre, Gre; Ana, Cyp].

DESCRIPTION. *Ophrys fusca* Link subsp. *cinereophila* (Paulus & Gack) N. Faurholdt
Syn. *O. cinereophila* Paulus & Gack.

FIG. 37B

Plant slender, 7–20(–25) cm tall with 2–8(–11) spirally arranged flowers in a relatively dense spike on the uppermost one third (or more) of the stem. **Sepals** (pale) green, 7–11.5 × 3–6 mm. **Petals** (olive–)green, often suffused with brown, 4–8 × 1–3 mm. **Lip** (pale) green underneath, with a moderate knee-like bend at the base, in cross section vaulted, 7–12 × 7–9 mm, with a narrow to relatively broad yellow or yellowish green margin; side lobes downcurved; **mirror** somewhat obscure, dull bluish to greyish or brownish violet. FIG. 37.

REFERENCES: Baumann & Künkele (1982, 1988), Berger (2003), Bernardos et al. (2005), Bournérias (1998), Buttler (1986), E. Danesch & O. Danesch (1969), Davies et al. (1983), Del Prete & Tosi (1988), Delforge (1994, 2001, 2005), Grasso & Manca (2002), Kullenberg (1961, 1973), Lorella et al. (2002), Nelson (1962), Paulus (1988, 1998, 2000, 2001, 2001b), Paulus & Gack (1980, 1981, 1983, 1986, 1990a, 1990b, 1990c, 1992, 1995), Rossi (2002), Souche (2004), Sundermann (1980), Vöth (1985).

4. *Ophrys lutea*

Just like *O. fusca*, *O. lutea* is one of the commonest bee orchids in the Mediterranean. Its variation patterns are not nearly as complex and confusing as those encountered in the former – but hardly any other *Ophrys* species exhibits a corresponding magnitude of variation in general appearance. Thus, *O. lutea* subsp. *melena* is quite humble with its small, predominantly green and brown flowers; the likewise small-flowered subsp. *galilaea* is bashful but pretty with considerably more yellow on the lip; finally, the large-flowered subsp. *lutea* is a marvellous plant with a broad yellow margin and contrasting dark central part of the lip. Certain forms of *O. lutea* may be confused with *O. fusca*, but they are consistently distinguished by the spreading to upcurved margin of the lip. Furthermore, the lip appears predominantly brown in *O. fusca*, while it appears predominantly yellow in *O. lutea* (except in subsp. *melena*).

   The morphological variation in *O. lutea* mainly falls between three groups of populations: (1) a large group that is dominant in the western part of the range; (2) a large group dominant in the eastern part; (3) a smaller group that is represented in Greece and Albania only. Each of these entities are pollinated by one or more andrenid bees of the genus *Andrena*, but their respective selections of pollinators appear not to overlap. Against this background we have decided to recognise the three groups of populations as separate subspecies, viz. subsp. *lutea*, subsp. *galilaea* and subsp. *melena*, respectively.

HABITAT. Dry to moist, preferably calcareous soil in full sunlight or light shade, from sea level to 1800 m altitude. Typical habitats include roadside slopes, garrigue, maquis, grassland, light-open pine and oak woods, as well as pesticide-free olive-groves.
FLOWERING TIME. From January to May, with a peak in March-April. In some years, flowering individuals may be found by the coast as early as December.
DISTRIBUTION. Throughout the Mediterranean and in western France north to Brittany. [Aeg, Alb, Bal, Cor, Cre, Fra, Gre, Ita, Mal, Por, Sar, Sic, Spa, Tur, Yug; Ana, Cyp, Isr, Mor, Tun] MAP 6.
DESCRIPTION. *Ophrys lutea* Cav.
Synonyms are listed under the subspecies.

Plant compact to slender, 5-40 cm tall with 1-6(-10) flowers in a relatively dense to lax spike. **Sepals** (pale) green to yellowish green, (ob)ovate to

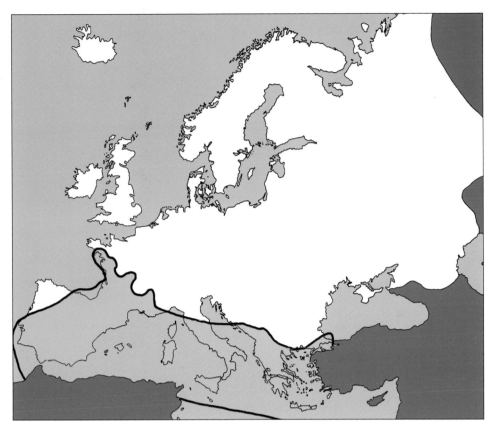

MAP 6. The European range of *Ophrys lutea*. Three subspecies are recognized in Europe; only in Albania and Greece do the geographic ranges of all three subspecies overlap.

elliptic, 6-13 × 3-10 mm; dorsal sepal more or less boat-shaped, incurved, from the base nearly parallel to the column. **Petals** yellowish green, (narrowly) oblong with flat margins, (3-)4.5-8 × 1-3 mm, glabrous, spreading to incurved. **Lip** with brown ground colour and an often very broad yellow margin, straight or with a knee-like bend at base, with flat-spreading to upcurved margins, distinctly longitudinally furrowed at the base, three-lobed close to the apex, 8-18 × 5-19 mm, velvety except on the yellow margin (and sometimes the mirror); bulges absent; mid-lobe distinctly longer than the side lobes, emarginate; **mirror** distinct to somewhat obscure, consisting of a figure shaped more or less like the tail of a fish (sometimes longitudinally split) and extending from the base of the lip to above the middle, dull greyish blue to dark greyish brown. **Column** rounded, not tapering towards the base (in side view); stigmatic cavity usually wider than long and approximately twice as wide as the anther, devoid of lateral, eye-like knobs at base. FIGS 38-40.

FIG. 38A

FIG. 38B

Key to the subspecies

1. Lip (13-)14-19 mm wide, with a distinct knee-like bend at base; the yellow margin of the side lobes 3-6 mm wide . . . . 4.1. **subsp.** *lutea*
1. Lip 5-10(-12) mm wide, straight (to slightly downcurved); the yellow margin of the side lobes up to 3 mm wide (rarely missing) 2
2. Side lobes of the lip predominantly yellow (the yellow margin 2-3 mm wide) . . . . 4.2. **subsp.** *galilaea*
2. Side lobes of the lip predominantly brownish orange to blackish brown (the yellow margin up to 1 mm wide, occasionally missing) . . . . 4.3. **subsp.** *melena*

4.1. *Ophrys lutea* subsp. *lutea*

Subsp. *lutea* is easily recognised in the western part of the Mediterranean. In that area, the broad yellow margin of the lip is very prominent, and individuals with brown markings on the mid-lobe are rare. However, mixed colonies with subsp. *galilaea* may be encountered locally (not least in Italy), and in such colonies it is often difficult to draw a firm line between the subspecies. Towards the eastern end of the Mediterranean, subsp. *lutea* is much rarer and even absent from certain regions. Eastern individuals of subsp. *lutea* have, in general, somewhat smaller flowers with a narrower yellow lip margin and more often with brown markings on the mid-lobe. Flowering

takes place from February to May, being at its best in March–April.

The geographic range encompasses most of the range of the species, east to Crete and Rhodes. Subsp. *lutea* is common in the western part of the area, particularly in Spain, Portugal and northern Africa, east to Tunisia. It is, however, rare in Sardinia and the Balearic Islands. [Aeg, Alb, Bal, Cor, Cre, Fra, Gre, Ita, Mal, Por, Sar, Sic, Spa, Yug; Ana, Mor, Tun].

DESCRIPTION. *Ophrys lutea* Cav. subsp. *lutea*
Syn. *O. phryganae* J. Devillers-Terschuren & P. Devillers.

Plant (7-)10-30(-40) cm tall with 1-6(-10) flowers. **Sepals** (yellowish) green, 10-13 × 6-10 mm. **Petals** 6-8 × 2-3 mm. **Lip** with a knee-like bend at base, 14-18 × (13-)14-19 mm; side lobes predominantly yellow (the yellow margin 3-6 mm broad); mid-lobe yellow (less frequently with brown markings), rounded to slightly emarginate; **mirror** distinct, dull greyish blue. FIG. 38.

4.2. *Ophrys lutea* subsp. *galilaea*

This subspecies is normally easy to identify, although mixed colonies with subsp. *lutea* in the central and eastern parts of the range may cause confusion (see comments under the latter). Furthermore, the delimitation towards subsp. *melena* is in need of further clarification (see under the latter). Flowering takes place from January (in eastern lowland areas) to June (in

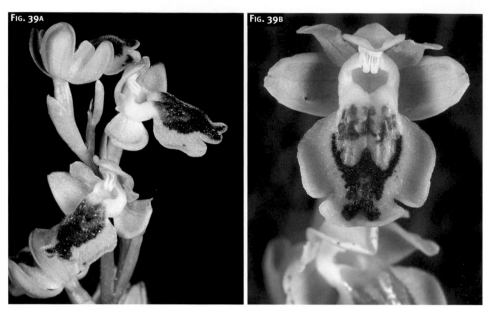

FIG. 39. *Ophrys lutea* subsp. *galilaea* (**A.** Greece, Chios, Elinda, 14th April 2005; **B.** Greece, Samos, Mykali, 12th April 1999). Photos by H. Æ. Pedersen.

mountains), in most regions with a distinct peak in March–April.

The geographic range encompasses most of the Mediterranean. However, subsp. *galilaea* is absent from France and extremely rare in the Iberian Peninsula. Thus, from mainland Spain we are only aware of our own two finds from Andalusia (both from the neighbourhood of Ronda, 1994), and from Portugal we have noticed no other finds than our own from the Algarve (between Malhao and Paderne, 1999). [Aeg, Alb, Cor, Cre, Gre, Ita, Mal, Por, Sar, Sic, Spa, Tur, Yug; Ana, Cyp, Isr, Mor, Tun].

DESCRIPTION. *Ophrys lutea* Cav. subsp. *galilaea* (H. Fleischm. & Bornm.) Soó Syn. *O. archimedea* P. Delforge & M. Walravens; *O. galilaea* H. Fleischm. & Bornm.; *O. lutea* Cav. subsp. *minor* (Tod.) O. & E. Danesch; *O. lutea* Cav. var. *minor* (Tod.) Guss.; *O. lutea* Cav. subsp. *murbeckii* sensu H. Sund. [non (H. Fleischm.) Soó?]; *O. numida* J. Devillers-Terschuren & P. Devillers; *O. sicula* Tineo.

Plant 5-25 cm tall with 2-6(-10) flowers. **Sepals** light green, 6-11 × 3-7 mm. **Petals** (3-)4.5-7.5 × 1-2.5 mm. **Lip** straight (rarely with a knee–like bend at base), 8-14.5 × 5-10(-12) mm; side lobes predominantly yellow (the yellow margin 2-3 mm broad, very rarely narrower); mid-lobe yellow with a brown figure shaped as a swallow's tail, more or less emarginate; **mirror** distinct, dull greyish blue. FIG. 39.

4.3 *Ophrys lutea* subsp. *melena*

Due to its predominantly brown lip, this subspecies has a superficial similarity to certain forms of *O. fusca* subsp. *fusca*, but it is distinguished by the spreading margins of the lip side lobes. In the central Mediterranean, subsp. *melena* has been reported from Italy (Monte Gargano and Sicily). However, the plants in question, as well as similar plants from western Crete, cannot be unequivocally referred to this subspecies and more likely represent local forms of subsp. *galilaea*. Where the two subspecies grow together in Greece, subsp. *melena* starts to flower a couple of weeks later than subsp. *galilaea*. Flowering starts in March and peaks in April.

Subsp. *melena* is distributed in the southern Balkans (north to Thessaloniki and southern Albania) as well as in Corfu and eastern Crete. Records from Lesbos, Tunisia, Algeria and Morocco are highly doubtful and probably represent confusion with hybrids between *O. fusca* subsp. *fusca* and *O. lutea* s.l. [Alb, Cre, Gre].

DESCRIPTION. *Ophrys lutea* Cav. subsp. *melena* Renz
Syn. *O. melena* (Renz) Paulus & Gack; *O. lutea* Cav. var. *melena* (Renz) E. Nelson.

FIG. 40

FIG. 40. *Ophrys lutea* subsp. *melena* (Greece, Attika, Imettos near Athens, 15th April 2000). Photo by N. Faurholdt.

Plant 10-40 cm tall with 2-6 flowers. **Sepals** (light) green, 7.5-11 × 4.5-7 mm. **Petals** 5.5-7.5 × 1.5-2.5 mm. **Lip** (nearly) straight, 10-14.5 × 7-10(-12) mm; side lobes predominantly brownish orange to blackish brown (the yellow margin up to 1 mm broad, sometimes absent); mid-lobe light brownish orange to blackish brown with or without an up to 1 mm broad yellow margin and sometimes with a darker figure shaped as a swallow's tail, more or less emarginate; **mirror** obscure, dark greyish brown. FIG. 40.

REFERENCES: Baumann & Künkele (1982, 1988), Bournérias (1998), Buttler (1986), E. Danesch & O. Danesch (1969), Davies et al. (1983), Del Prete & Tosi (1988), Delforge (1994, 2001, 2005), Grasso & Manca (2002), Kullenberg (1961, 1973), Nelson (1962), Paulus (1988, 2001b), Paulus & Gack (1980, 1986, 1990a, 1990b, 1992, 1995), Rossi (2002), Souche (2004), Sundermann (1980), Vöth (1984, 1985).

5. *Ophrys tenthredinifera*

Ophrys tenthredinifera is a gem among bee orchids. Its flowers are large and gaily coloured, dominated by shades of rose, yellow and dark brown, and many people consider this species the prettiest member of the genus. It is easy to identify and can hardly be mistaken for other species. The pollinators of O. *tenthredinifera* include a number of anthophorid bees, mainly of the genus *Eucera*.

The appearance of the flowers varies a great deal. This has led to the description of a number of subspecies and varieties, and recently even to splitting at the species level. In our opinion, however, the variation does not support a systematic (sub)division of the species. Particularly in eastern

FIG. 41A

FIG. 41. *Ophrys tenthredinifera* (**A.** Italy, Sardinia, Osini, 20th April 1997; **B.** Spain, Andalusia, Grazalema, 18th April 2001; **C.** Italy, Sicily, Buccheri, 17th April 1999; **D.** Spain, Balearic Islands, Mallorca, Valldemosa, 18th April 1990; **E.** Greece, Chios, Tavros, 15th April 2005). Photos by H. Æ. Pedersen (**A-C, E**), N. Faurholdt (**D**).

FIG. 41B

FIG. 41C

FIG. 41D

FIG. 41E

Mediterranean individuals, the lip is sometimes smaller and with a darker greyish brown (not yellowish brown) ground colour. In Crete as well as in Chios, Samos and Tunisia we have seen this form growing side by side with "normal" individuals. Very early-flowering, but morphologically indistinguishable, populations can be encountered in central Portugal. In Mallorca, Sicily and mainland Spain (particularly Andalusia) one can find tall individuals with unusually large and colourful flowers, as well as individuals with reflexed sepals. An apochromic form with light yellowish green lip and whitish green sepals is occasionally encountered, most frequently in Crete. Finally, it should be mentioned that an apparent hybrid between *O. fuciflora* subsp. *candica* and *O. tenthredinifera* occurs in southern Italy (Puglia). This putative hybrid is often recognised as a distinct species, *O. tardans* O. & E. Danesch.

HABITAT. Calcareous to slightly acid soil, in full sunlight or light shade from sea level to 1800 m altitude. Typical habitats include roadside verges, grassland, garrigue, open pine woods and old olive groves.

FLOWERING TIME. From February to May, but with a distinct peak from mid-March to mid-April. In many lowland areas the flowering is over by the beginning of April.

DISTRIBUTION. Most of the Mediterranean, where it is fairly common in the western and central parts of the range, but its frequency decreases towards the east – for example, it is rare on several Aegean islands. In Cyprus, *O. tenthredinifera* is known from one locality only, and it has not been found in the Middle East. [Aeg, Alb, Bal, Cor, Cre, Fra, Gre, Ita, Mal, Por, Sar, Sic, Spa, Tur, Yug; Ana, Cyp, Mor, Tun] MAP 7.

DESCRIPTION. *Ophrys tenthredinifera* Willd.

Syn. *O. aprilia* P. Devillers & J. Devillers-Terschuren; *O. ficalhoana* J. A. Guim.; *O. tenthredinifera* Willd. var. *ficalhoana* J. A. Guim.; *O. grandiflora* Tenore; *O. tenthredinifera* Willd. subsp. *guimaraesii* D. Tyteca; *O. tenthredinifera* Willd. var. *mariana* Rivas Goday; *O. neglecta* Parl.; *O. tenthredinifera* Willd. subsp. *praecox* D. Tyteca, nom. illeg.; *O. tenthredinifera* Willd. var. *praecox* Rchb.f. ex E. G. Camus et al.; *O. tenthredinifera* Willd. var. *ronda* Schltr.; *O. tenthredinifera* Willd. subsp. *villosa* (Desf.) H. Baumann & Künkele.

Plant compact to slender, 10–30(-45) cm tall with (1-)2–10 flowers in a (relatively) dense spike. **Sepals** violet to white, (broadly) elliptic, 11–13 × 6.5–10 mm; dorsal sepal more or less boat-shaped, slightly incurved, from the base reflexed. **Petals** of the same colour as the sepals, (ovate-)triangular

MAP 7. The European range of *Ophrys tenthredinifera*. This species is not subdivided into subspecies.

(often auriculate) with flat margins, 4–6 × 2.5-6 mm, velvety to shaggy, spreading. **Lip** with brown ground colour and a broad yellow (to light brown) margin, straight with recurved sides, entire, 9–16 × 10-20 mm, more or less hairy (especially along the margin) and with a particularly prominent tuft of hairs close to the apex; bulges weakly developed, distinctly isolated from the margin of the lip; apex emarginate, provided with a short erect, narrowly triangular point; **mirror** distinct, consisting of a simple and fairly small H– to horseshoe-shaped figure, the basal arms of which are connected to the base of the lip, dull greyish blue to greyish violet with a pale border. **Column** rounded, not tapering towards the base (in side view); stigmatic cavity at least as wide as long and approximately twice as wide as the anther, devoid of lateral, eye-like knobs at base. FIG. 41.

REFERENCES: Baumann & Künkele (1982, 1988), Baumbach (2003), Bournérias (1998), Buttler (1986), E. Danesch & O. Danesch (1969), Davies et al. (1983), Del Prete & Tosi (1988), Delforge (1994, 2001, 2005), Devillers et al. (2003), Kullenberg (1961, 1973), Kullenberg, Borg-Karlson & Kullenberg (1984), Nelson (1962), Paulus (1988), Paulus & Gack (1980, 1981), Rossi (2002), Souche (2004), Sundermann (1980), Vöth (1984).

MAP 8. The distribution of *Ophrys bombyliflora* in Europe (outside the map, the Canary Islands are also encompassed by the European range). This species is not subdivided into subspecies.

6. *Ophrys bombyliflora*

With its small, bumblebee-like flowers O. *bombyliflora* is almost invariable and impossible to confuse with other species. The plants are usually very small, and only their habit of growing in large swarms with 5-10 cm between the flowering shoots prevents them from being easily overlooked. This gregarious way of growing is determined by the vegetative reproduction of this species, which is unique within the genus (see the section Annual growth cycle and vegetative reproduction p. 22). The flowers are pollinated by several anthophorid bees of the genus *Eucera*.

HABITAT. Dry to very wet, calcareous soil, in full sunlight to light shade from sea level to 900 m altitude. Typical habitats include garrigue, roadside slopes, grassland, open forest, pesticide-free olive groves and moist meadows.

FLOWERING TIME. From February to May, with a peak in March–April.

DISTRIBUTION. Throughout the Mediterranean, except for the very easternmost regions, and it is the only *Ophrys* to reach the Canary Islands in

the Atlantic. Close to its eastern boundary, this species is infrequent (e.g. on Rhodes) to very rare (e.g. on Lesbos and Samos). In many parts of the central Mediterranean, on the other hand, *O. bombyliflora* is common. [Aeg, Alb, Bal, Can, Cor, Cre, Fra, Gre, Ita, Mal, Por, Sar, Sic, Spa, Tur, Yug; Ana, Mor, Tun] MAP 8.

DESCRIPTION. *Ophrys bombyliflora* Link.

FIG. 42. *Ophrys bombyliflora* (**A.** Italy, Sardinia, Osini, 21st April 1997; **B.** Tunisia, Cap Bon, 16th March 1998; **C.** Greece, Samos, Zervos, 12th April 1999). Photos by H. Æ. Pedersen.

Plant compact, 5-20(-30) cm tall with 1-5 flowers in a relatively dense spike. **Sepals** (yellowish) green, broadly elliptic, 9-12 × 6-8 mm; dorsal sepal more or less boat-shaped, slightly incurved, from the base reflexed. **Petals** (yellowish) green, often suffused with brown towards the base, (triangular-)ovate with flat margins, 3-4 × 3-4 mm, velvety to shaggy, spreading. **Lip** with (greyish) brown ground colour, straight with strongly recurved sides, deeply three-lobed in its basal part, 8-10 × 11-13 mm, velvety to shaggy along the margin (otherwise nearly glabrous); side lobes converted into obliquely conical bulges; mid-lobe longer than the side lobes, emarginate, provided with a short, porrect to downward pointing, awl-shaped point (hidden under the strongly vaulted lip); **mirror** obscure, consisting of a simple and relatively small, somewhat H- to horseshoe-shaped figure, the basal arms of which are connected to the base of the lip, dull (violet-)grey, occasionally with a pale border. **Column** rounded, not tapering towards the base (in side view); stigmatic cavity at least as wide as long and approximately twice as wide as the anther, devoid of lateral, eye-like knobs at base. FIG. 42.

REFERENCES: Baumann & Künkele (1982, 1988), Bournérias (1998), Buttler (1986), E. Danesch & O. Danesch (1969), Davies et al. (1983), Del Prete & Tosi (1988), Delforge (1994, 2001, 2005), Gulli et al. (2003), Kullenberg (1961, 1973), Nelson (1962), Paulus (1988), Rossi (2002), Souche (2004), Sundermann (1980).

7. *Ophrys speculum*

"Speculum" is the Latin word for mirror, and the name undoubtedly refers to the fact that this species has the largest and most shining mirror of any *Ophrys*. It is generally one of the shortest bee orchids, but it is also one of the commonest in the major part of the Mediterranean and often forms extensive populations. Especially in large populations of O. *speculum* subsp. *speculum* there is a good chance to observe the scoliid wasp *Dasyscolia ciliata* visit and pollinate the flowers (the pollinators of the two other subspecies have not yet been recorded). *Ophrys speculum* is the only bee orchid to be pollinated by wasps of the family Scoliidae.

Because of its large, metallic blue mirror, O. *speculum* can hardly be confused with other species but, nevertheless, exhibits considerable variation in morphological characters. Most of the variation falls between three groups of populations which are additionally limited to partly different geographic regions and show partly differing flowering times. Based on these features, we recognise them as three subspecies.

HABITAT. Dry to moist, calcareous soil in full sunlight or light shade from sea level to 1200 m altitude. Typical habitats include garrigue, poor grassland, roadside slopes, open pine woods, fallow fields and pesticide-free olive groves.

FLOWERING TIME. From late February to early May, with a peak from mid-March to mid-April.

DISTRIBUTION. This species is almost universally distributed in the Mediterranean. It seems absent from Cyprus and is very rare in France and Crete – indeed, it has been recorded only a couple of times from Crete. [Aeg, Alb, Bal, Cor, Cre, Fra, Gre, Ita, Mal, Por, Sar, Sic, Spa, Tur; Ana, Mor, Tun] MAP 9.

DESCRIPTION. *Ophrys speculum* Link, nom. cons.

Synonyms are listed under the subspecies.

Plant compact to slender, 5-50 cm tall with 2-15 flowers in a lax spike. **Sepals** dark green to pale green, often with reddish brown to purplish brown streaks, spots, or suffusion, elliptic to oblong, 6-10 × 3-5 mm; dorsal

MAP 9. The European range of *Ophrys speculum*. Three subspecies are recognized in Europe; the widespread subsp. *speculum* is accompanied by subsp. *lusitanica* in southern Spain and Portugal and by subsp. *regis-ferdinandii* in the eastern Aegean Islands.

sepal distinctly boat-shaped, strongly incurved, from the base nearly parallel to the column. **Petals** yellowish green, weakly to strongly suffused with brown or red, elliptic to triangular-ovate with flat margins, 4-6 × 4-5 mm, velvety, recurved. **Lip** with light brown to (yellowish) green ground colour around the large mirror, straight with spreading to recurved sides, deeply three-lobed close the middle, 11-16 × 8-10 mm, strongly shaggy by a dense reddish brown hairiness along the margin (otherwise nearly glabrous); bulges absent, mid-lobe distinctly longer than the side lobes, emarginate; **mirror** distinct, covering almost the entire mid-lobe as one large patch, shining blue. **Column** rounded, not tapering towards the base (in side view); stigmatic cavity approximately as wide as long and approximately as wide as the anther, with lateral, dark, eye-like knobs at base. FIGS 43-45.

Key to the subspecies

1. Neither the mid-lobe nor the side lobes of the lip with distinctly recurved margins; side lobes obliquely oblong (0.8-1.5 times as long as wide), slightly inclined . . . .
. . . . 7.1. **subsp. *speculum***
1. The mid-lobe as well as the side lobes of the lip with distinctly recurved margins; side lobes obliquely (linear-) oblong (more than 1.5 times as long as wide), straight 2

2. Lateral sepals pale green, only weakly (if at all) suffused with violet. Mid-lobe of the lip 1-2.2 times as long as wide . . . .
. . . . .7.2. **subsp. *lusitanica***
2. Lateral sepals strongly suffused with brownish violet. Mid-lobe of the lip 2-4 times as long as wide . . . .
. . . . 7.3. **subsp. *regis-ferdinandii***

FIG. 43A

FIG. 43. *Ophrys speculum* subsp. *speculum* (**A.** Italy, Sicily, Necropoli di Pantalica, 18th April 1999; **B.** Italy, Sardinia, Ponte su Crabiolu, 20th April 1997; **C.** Italy, Sardinia, Tertenia, 23rd April 1997). Photos by H. Æ. Pedersen.

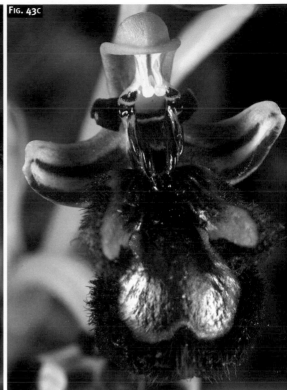

7.1. *Ophrys speculum* subsp. *speculum*

This is by far the commonest and most widely distributed subspecies.
Flowering takes place from February to April, with a distinct peak around
the start of April.

Subsp. *speculum* is distributed throughout the range of the species. It is
very rare in France and Crete. Records from Cyprus are probably
erroneous. [Aeg, Alb, Bal, Cor, Cre, Fra, Gre, Ita, Mal, Por, Sar, Sic, Spa,
Tur; Ana, Mor, Tun].

DESCRIPTION. *Ophrys speculum* Link subsp. *speculum*
Syn. *O. ciliata* Biv.; *O. vernixia* Brot. subsp. *ciliata* (Biv.) Del Prete; *O.
vernixia* Brot. subsp. *orientalis* Paulus.

Plant 5-25(-30) cm tall with 2-8 flowers. Lateral *sepals* (dark green), more
or less streaked or spotted with brownish violet. **Petals** dark brownish
violet. **Lip** with spreading to slightly recurved margin; side lobes obliquely
oblong, 0.8-1.5 times as long as wide, slightly inclined, mid-lobe (broadly)
obovate, 0.6-1.1 time as long as wide. FIG. 43.

FIG. 44. *Ophrys speculum* subsp. *lusitanica* (**A.** Portugal, the Algarve, Paderne, 4th April 1998; **B.** Portugal, the Algarve, Malhao, 31st March 1999). Photos by N. Faurholdt.

7.2. *Ophrys speculum* subsp. *lusitanica*

Subsp. *lusitanica* can be easily recognised by the features indicated in the key. Variation is modest, and potentially confusing hybrids with subsp. *speculum* are not seen very often. Flowering takes place from March to May, beginning a couple of weeks later than in subsp. *speculum*.

The distribution of this subspecies is western and fragmented. Today, subsp. *lusitanica* is known from three distinct areas – two in Portugal (the Algarve and Estremadura/Ribatejo/Beira) and one in Spain (Andalusia). Even within these areas, the plant occurs only locally. [Por, Spa].

FIG. 44B

DESCRIPTION. *Ophrys speculum* Link subsp. *lusitanica* O. & E. Danesch Syn. *O. vernixia* Brot. s.s.; *O. vernixia* Brot. subsp. *lusitanica* (O. & E. Danesch) H. Baumann & Künkele.

Plant 15-50 cm tall with 5-15 flowers. Lateral *sepals* only weakly (if at all) suffused with violet. **Petals** light green or orange. **Lip** with strongly recurved margin; side lobes obliquely (linear-)oblong, 1.5-3.5 times as long as wide, straight and divergent; mid-lobe (narrowly) obovate, 1-2.2 times as long as wide. FIG. 44.

7.3 *Ophrys speculum* subsp. *regis-ferdinandii*

Subsp. *regis-ferdinandii* is well delimited and easy to recognise. Hybrids with subsp. *speculum* are relatively rare, but they have been found in Rhodes and a few other places. Flowering takes place in March–April, beginning a couple of weeks later than in subsp. *speculum*.

Subsp. *regis-ferdinandii* is endemic to an area encompassing some of the easternmost Aegean islands (viz. Chios, Samos, Tilos, Sini and Rhodes) and immediately neighbouring parts of southwestern Anatolia. On Chios and Rhodes, this subspecies is widespread and fairly common. On Samos, on the other hand, it is very rare and only found in the southwestern part of the island. [Aeg; Ana].

FIG. 45. *Ophrys speculum* subsp. *regis-ferdinandii* (**A.** Greece, Rhodes, Katavia, 26th March 2002; **B.** Greece, Chios, Elinda, 14th April 2005). Photos by N. Faurholdt (**A**), H. Æ. Pedersen (**B**).

DESCRIPTION. *Ophrys speculum* Link subsp. *regis-ferdinandii* Achtaroff & Kellerer ex Kuzmanov
Syn. *O. regis-ferdinandii* (Achtaroff & Kellerer ex Renz) Buttler; *O. speculum* Link var. *regis-ferdinandii* (Achtaroff & Kellerer ex Renz) Soó; *O. vernixia* Brot. subsp. *regis-ferdinandii* (Achtaroff & Kellerer ex Kuzmanov) Renz & Taubenheim.

Plant 5–30 cm tall with 2–11 flowers. Lateral **sepals** green, strongly suffused with violet to reddish brown. **Petals** usually a little paler, violet to brown. **Lip** with strongly recurved margin; side lobes obliquely linear-oblong, 2–4.5 times as long as wide, straight and divergent, mid-lobe obovate-oblanceolate, 2–4 times as long as wide. Fig. 45.

REFERENCES: Ayasse et al. (2003), Baumann & Künkele (1982, 1988), Bournérias (1998), Buttler (1983, 1986), E. Danesch & O., Danesch (1969), Davies et al. (1983), Del Prete & Tosi (1988), Delforge (1994, 2001, 2005), Greuter (2004), Hansen et al. (1990), Jacobsen & Rasmussen (1976), Kullenberg (1961, 1973), Nelson (1962), Paulus (2001b), Paulus & Gack (1980, 1981), Rossi (2002), Souche (2004), Sundermann (1980).

8. *Ophrys insectifera*

In many temperate parts of Europe, *O. insectifera* is the only occurring species of *Ophrys*, but it is also the only one to be almost entirely absent from the subtropical lowland of the Mediterranean. *Ophrys insectifera* is furthermore remarkable by having the indisputably most insect-like flowers in the whole genus. The lobed, hairy lip with the glabrous mirror mimics the body and the folded wings; two small, dark, shining knobs at the base of the lip represent the eyes; and the two thread-like petals are perfect antennae. Indeed, the appearance of the flowers is so characteristic that this species can hardly be confused with others.

Ophrys insectifera is not very variable, but the variation that it does exhibit mainly falls between two groups of populations. These groups have overlapping distributions, but different pollinators, for which reason we recognise them as two subspecies. Subsp. *insectifera* is pollinated by the digger-wasps *Argogorytes mystaceus* and *A. fargeii* (family Sphecidae) and, locally, by *Sterictiphora furcata* (family Argidae). Subsp. *aymoninii* is pollinated by the andrenid bee *Andrena combinata*. The Spanish local populations of subsp. *insectifera* that are pollinated by *Sterictiphora furcata* have recently been described as *O. subinsectifera* and might well deserve systematic recognition.

HABITAT. Calcareous, dry to wet soil in full sunlight to light shade from sea level to 1700 m altitude. Typical habitats include fens, grassland, open woods and wooded meadows.

FLOWERING TIME. From May to July.

DISTRIBUTION. Mainly central European, stretching from Ireland and the mountains of northern Spain east to the Apennines, Rumania, northern Greece and the Ukraine. In the northeast there are outliers to Norway and the Moscow area. [Alb, Aus, Bal, Bel, Cze, Den, Eng, Est, Fin, Fra, Ger, Gre, Hol, Hun, Ire, Ita, Lat, Lit, Lux, Nor, Pol, Rum, Rus, Spa, Swe, Swi, Ukr, Wal, Yug] MAP 10.

DESCRIPTION. *Ophrys insectifera* L.

Synonyms are listed under the subspecies.

Plant slender to compact, 15-50(-80) cm tall with 2-15(-20) flowers in a lax spike. **Sepals** pale green to yellowish green, elliptic to lanceolate-oblong, 6-9 × 3-5 mm; dorsal sepal nearly flat, straight, from the base nearly parallel to the column. **Petals** dark brown to (yellowish) green, linear with recurved margins, 4-7 × 1(-1.5) mm, velvety, more or less porrect. **Lip** with brown ground colour and occasionally a broad, green to yellow margin, straight with slightly recurved margin, three-lobed close to the middle, 9-15 × 6-12 mm, velvety except on the mirror; bulges absent; mid-lobe much longer than the side lobes, deeply cleft; **mirror** distinct, consisting of a central, rectangular to somewhat butterfly-shaped figure without connection to the base of the lip, dull bluish grey. **Column** rounded, not tapering towards the base (in side view); stigmatic cavity approximately as wide as long and approximately as wide as the anther, devoid of lateral, eye-like knobs at base. FIGS 46-47.

Key to the subspecies

1. Petals dark brown. Lip longer than wide; mid-lobe and side lobes brown (very rarely with green to yellowish margins). Column with reddish anther locules . . . . 8.1. **subsp.** *insectifera*
1. Petals (yellowish) green (sometimes weakly suffused with brown). Lip approximately as long as wide; mid-lobe and side lobes brown with broad yellow margins. Column with yellow anther locules . . . . 8.2. **subsp.** *aymoninii*

8.1. *Ophrys insectifera* subsp. *insectifera*

This subspecies is the more variable of the two. Forms with yellowish margins of the lip are reminiscent of subsp. *aymoninii*, but they are distinguished from the latter in other characters (cf. the key). The habitats

FIG. 46A

FIG. 46. *Ophrys insectifera* subsp. *insectifera* (**A.** Denmark, Zealand, Allindelille Fredskov, 5th June 1999; **B,D.** Sweden, Öland, Runsten, 9th June 2000; **C.** Estonia, Saaremaa, Jaagarahu, 21st June 2001). Photos by H. Æ. Pedersen (**A,C**), N. Faurholdt (**B,D**).

FIG. 46B

FIG. 46C

FIG. 46D

113

FIG. 47. *Ophrys insectifera* subsp. *aymoninii* (**A-B.** France, Massif Central, Lanuejols, 18th May 2002). Photos by B. T. Christensen.

are the same as indicated under the species; in northern Europe subsp. *insectifera* mainly occurs in calcareous fens. Flowering takes place from April to June; in northern Europe it starts during the second half of May and peaks by mid-June, whereas in central and southern Europe, the flowers start to open by mid-April.

Subsp. *insectifera* is distributed throughout the range of the species. [Alb, Aus, Bal, Bel, Cze, Den, Eng, Est, Fin, Fra, Ger, Gre, Hol, Hun, Ire, Ita, Lat, Lit, Lux, Nor, Pol, Rum, Rus, Spa, Swe, Swi, Ukr, Wal, Yug].

DESCRIPTION. *Ophrys insectifera* L. subsp. *insectifera*
Syn. *O. subinsectifera* Hermosilla & Sabando.

Plant slender, 15-50(-80) cm tall; the vegetative parts dull (bluish) green. *Spike* with 2-15(-20) flowers. **Petals** dark brown. **Lip** longer than wide, 9-15 × 6-10 mm; mid-lobe and side lobes brown (very rarely with green to yellowish margins). **Column** with reddish anther locules. FIG. 46.

8.2. *Ophrys insectifera* subsp. *aymoninii*

This subspecies is easily recognised and little variable. It grows in calcareous, dry to humid soil in open (beech/)pine woods as well as on grassland and edges of woods, at 500–1000 m altitude. Flowering takes place in May–June.

Subsp. *aymoninii* is endemic to France, where it is known from a small area in the Massif Central. An isolated occurrence in Alsace may not be of natural origin. [Fra].

DESCRIPTION. *Ophrys insectifera* L. subsp. *aymoninii* Breistr.
Syn. *O. aymoninii* (Breistr.) Buttler.

Plant slender to compact, 20–50 cm tall; the vegetative parts yellowish green. *Spike* with 5–10 flowers. **Petals** (yellowish) green (sometimes weakly suffused with brown). **Lip** approximately as long as wide, 9–12 × 8–12 mm; mid-lobe and side lobes brown with broad yellow margins. **Column** with yellow anther locules. FIG. 47.

REFERENCES: Ågren & Borg-Karlson (1984), Baumann & Künkele (1982, 1988), Bjørndalen (2006), Bournérias (1998), Buttler (1986), E. Danesch & O. Danesch (1969), Davies et al. (1983), Del Prete & Tosi (1988), Delforge (1994, 2001, 2005), Engel & Mathé (2002), Hermosilla et al. (1999), Kullenberg (1961, 1973), Nelson (1962), Rossi (2002), Souche (2004), Sundermann (1980), Wolff (1950).

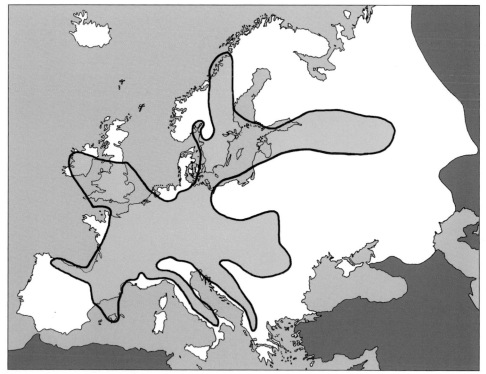

MAP 10. The total range of *Ophrys insectifera*. Two subspecies are recognized; subsp. *insectifera* is widespread, whereas subsp. *aymoninii* appears endemic to Massif Central, France.

9. *Ophrys apifera*

This stout and handsome *Ophrys* is encountered throughout most of western and southern Europe. It is recognised at once because of the extended and sigmoidly curved apex of the column, and its unique self-pollination can easily be observed (the pollinia have pivoted forwards and downwards, and the pollen masses now rest on the receptive part of the stigma, while the flaccid stalks of the pollinia form a bow). The self-pollination in *O. apifera*, which is further described in the section From

FIG. 48. *Ophrys apifera* (**A.** Italy, Sardinia, Tertenia, 23rd April 1997; **B.** Wales, Merioneth, Morfa Harlech, 14th June 1999; **C-D.** Italy, Sardinia, Ponte su Crabiolu, 20th April 1997). Photos by H. Æ. Pedersen (**A-B,** **D**), N. Faurholdt (**C**).

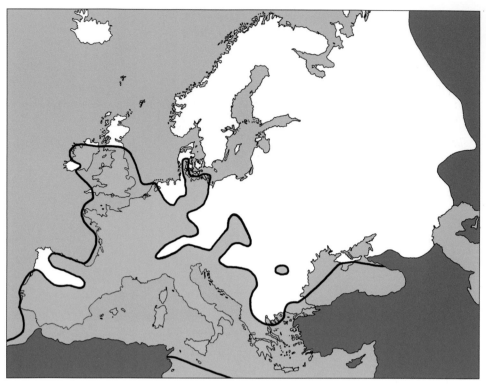

MAP 11. The European range of *Ophrys apifera*. This species is not subdivided into subspecies.

pollination to seed dispersal p. 24, must be considered one of the preconditions for the strikingly high frequency of monstrous forms in this species (see the section Monstrous forms p. 18). These forms (see FIG. 5 for a few examples) are fascinating, but because of their casual nature, they do not deserve to be recognised at varietal level or above.

HABITAT. Dry to wet calcareous soil in full sunlight to light shade, from sea level to 1800 m altitude. Typical habitats include edges of woods, open deciduous forest and pine woods, garrigue and grassland as well as open reed swamps and stabilised coastal dunes. Additionally, it is frequently seen as a pioneer plant in places such as recently abandoned quarries, railway embankments and roadside verges.

FLOWERING TIME. Flowering is relatively late, generally taking place from April to July. At warm coastal sites in the eastern Mediterranean, however, flowering individuals are regularly encountered as early as March. The peak flowering time is from April–June.

DISTRIBUTION. Throughout the Mediterranean, including the Levant, and in Atlantic western Europe north to the British Isles, Holland and Denmark.

[Aeg, Alb, Aus, Bal, Bel, Bul, Cor, Cre, Cze, Den, Eng, Fra, Ger, Gre, Hol, Hun, Ire, Ita, Lux, Mal, Por, Rum, Sar, Sco, Sic, Spa, Swi, Tur, Ukr, Wal, Yug; Ana, Cyp, Isr, Mor, Tun] MAP 11.

DESCRIPTION. *Ophrys apifera* Huds.

Syn. *O. apifera* Huds. var. *aurita* Moggr.; *O. apifera* Huds. var. *bicolor* (Nägeli) E. Nelson; *O. apifera* Huds. var. *botteronii* (Chodat) Asch. & Graebn.; *O. apifera* Huds. var. *chlorantha* (Hegetschw.) Arcang. [non (Hegetschw.) K. Richt., comb. superfl.]; *O. apifera* Huds. subsp. *jurana* Ruppert var. *friburgensis* (Freyhold) Ruppert; *O. apifera* Huds. var. *fulvofusca* M. P. Grasso & Scrugli; *O. holoserica* (Burm.f.) Greuter s.s., nom. tant.; *O. apifera* Huds. subsp. *jurana* Ruppert s.s.; *O. apifera* Huds. var. *trollii* (Hegetschw.) Rchb.f. [non (Hegetschw.) E. Nelson, comb. superfl.].

Plant slender, (15-)20-50(-70) cm tall with (2-)3-12(-17) flowers in a (relatively) lax spike. **Sepals** violet to white, narrowly ovate to lanceolate-oblong, 11-17 × 5-9 mm; dorsal sepal boat-shaped, straight to slightly incurved, from the base reflexed. **Petals** yellowish green to rose-coloured, sometimes suffused with red, triangular (often auriculate) with recurved margins, 1-3(-7) × 1(-2) mm, shaggy, spreading. **Lip** with (dark) brown ground colour, straight with strongly recurved margin, deeply (to moderately) three-lobed close to the base, 8-14 × 10-16 mm, velvety to shaggy along the margin (otherwise nearly glabrous); side lobes converted into obliquely conical bulges; mid–lobe much longer than the side lobes, rounded, provided with a downward pointing, more or less rectangular to rhomboid appendage (often hidden under the strongly vaulted lip); **mirror** distinct (rarely obscure), consisting of a more or less H-shaped or often slightly more complicated figure, the basal arms of which are connected to the base of the lip (occasionally, there are also a few isolated markings towards the apex), dull greyish violet with a cream border. **Column** extended into a sigmoidly curved apex, not tapering towards the base (in side view); stigmatic cavity approximately as wide as long and approximately twice as wide as the anther, with dark, lateral, eye-like knobs at base; pollinia with flaccid stalks. FIG. 48.

REFERENCES: Baumann & Künkele (1982, 1988), Bournérias (1998), Buttler (1986), E. Danesch & O. Danesch (1969), Davies et al. (1983), Del Prete & Tosi (1988), Delforge (1994, 2001, 2005), Gerasimova et al. (1998), Nelson (1962), Petersen & Høyer-Nielsen (2005), Rossi (2002), Souche (2004), Sundermann (1980).

10. *Ophrys umbilicata*

Ophrys umbilicata is mainly distributed in the Levant, but it also stretches across Greece to Albania. It is strongly reminiscent of O. *scolopax* (which is more widespread in Europe) but distinguished especially by the fact that its incurved dorsal sepal forms a roof over the column.

Within Europe, the morphological variation mainly falls between two groups of populations. These groups have overlapping distributions, but apparently different pollinators (and somewhat divergent flowering times), for which reason we recognise them as two subspecies. Subsp. *umbilicata* is pollinated by various anthophorid bees of the genus *Eucera*. Subsp. *bucephala*, on the other hand, appears to be pollinated by *Eucera curvitarsis* only – a species not reported as pollinator of subsp. *umbilicata*.

HABITAT. Dry to somewhat moist, often calcareous soil in full sunlight to light shade, from sea level to 1200 m altitude. Typical habitats include roadside slopes, grassland, garrigue, light–open pine woods and cypress groves as well as pesticide-free olive groves.

FLOWERING TIME. From February to April, with a peak in late March and early April.

DISTRIBUTION. From Albania through Greece (missing in Crete) to the Levant. [Aeg, Alb, Gre, Tur; Ana, Cyp, Isr] MAP 12.

DESCRIPTION. *Ophrys umbilicata* Desf.

Synonyms are listed under the subspecies.

Plant relatively slender to compact, 10-45(-60) cm tall with 2-12 flowers in a relatively lax to dense spike. **Sepals** rose-coloured to white or green, sometimes suffused with violet, (ovate-)elliptic, 8-17.5 × 4-9 mm; dorsal sepal more or less boat-shaped, strongly incurved, from the base nearly parallel to the column. **Petals** (yellowish) green to rose-coloured, sometimes suffused with red, (oblong-)triangular (often auriculate) with nearly flat margins, 3.5-8 × 2-5 mm, velvety, recurved. **Lip** with (reddish) brown to dark brown ground colour, straight with strongly recurved sides, deeply three-lobed close to its base, 7-15.5 × 8.5-19 mm, brownish shaggy on the outer side of the side lobes and velvety along the margin of the mid-lobe (otherwise nearly glabrous); side lobes converted into obliquely conical (to horn-like) bulges; mid-lobe longer than the side lobes, obtuse to emarginate, provided with a broad and conspicuous, erect (to porrect), rectangular, rhomboid or (ob)triangular, often dentate appendage; **mirror** distinct, consisting of a complicated (or rarely somewhat H-shaped) figure, the basal

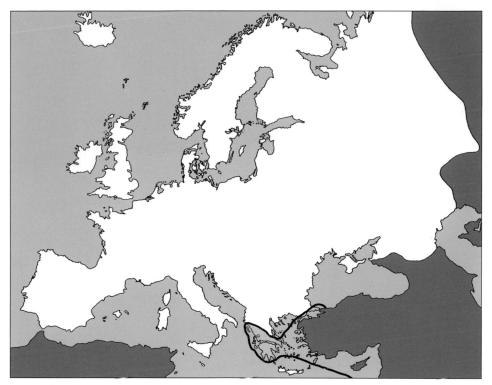

Map 12. The European range of *Ophrys umbilicata*. Two subspecies are recognized in Europe; subsp. *bucephala* has its only known European populations on Lesbos, Chios and Samos in the Aegean Sea, whereas subsp. *umbilicata* occurs throughout the range of the species.

arms of which are connected to the base of the lip, dull greyish blue to violet (more rarely reddish brown) with a cream border. **Column** acute to obtuse, not tapering towards the base (in side view); stigmatic cavity at least as wide as long and approximately twice as wide as the anther, with dark or pale, lateral, eye-like knobs at base. FIGS 49–50.

Key to the subspecies

1. Plant relatively slender (the uppermost foliage leaf hardly ever reaching past the node of the lowermost flower). Lip 8.5–15 mm wide . . . . 10.1. **subsp. *umbilicata***
1. Plant compact (the uppermost foliage leaf usually reaching past the node of the lowermost flower). Lip 15–19 mm wide . . . . 10.2. **subsp. *bucephala***

10.1. *Ophrys umbilicata* subsp. *umbilicata*

This subspecies exhibits a typical example of clinal variation. In the westernmost part of its range, almost all individuals have green sepals, but

FIG. 49. *Ophrys umbilicata* subsp. *umbilicata* (**A.** Greece, Samos, Agia Triada, 12th April 1999; **B-C.** Greece, Chios, Profitis Ilias, 15th April 2005; **D.** Greece, Attika, Imettos near Athens, 12th April 1998). Photos by H. Æ. Pedersen (**A-C**), N. Faurholdt (**D**).

FIG. 49B · FIG. 49C · FIG. 49D

further east they are gradually replaced by individuals with rose-coloured to white sepals, and in the Levant such plants are by far the most frequent. Subsp. *umbilicata* flowers from February to April.

The distribution encompasses the whole range of the species. [Aeg, Alb, Gre, Tur; Ana, Cyp, Isr].

DESCRIPTION. *Ophrys umbilicata* Desf. subsp. *umbilicata*
Syn. *O. attica* (Boiss. & Orph.) B. D. Jacks.; *O. carmeli* H. Fleischm. & Bornm. subsp. *attica* (Boiss. & Orph.) Renz; *O. scolopax* Cav. subsp. *attica* (Boiss. & Orph.) E. Nelson; *O. umbilicata* Desf. subsp. *attica* (Boiss. & Orph.) J. J. Wood; *O. carmeli* H. Fleischm. & Bornm. s.s.; *O. carmeli* H. Fleischm. & Bornm. subsp. *orientalis* (Renz) Soó; *O. scolopax* Cav. subsp. *orientalis* (Renz) E. Nelson.

Plant relatively slender (the uppermost foliage leaf hardly ever reaching past the node of the lowermost flower), 10-45(-60) cm tall with 2-12 flowers in a relatively lax spike. **Sepals** rose-coloured to green or white, 8-15 × 4-8 mm. **Petals** 3.5-6.5 × 2-4 mm. **Lip** 7-12.5 × 8.5-15 mm; side lobes with straight or outcurved apices. FIG. 49.

10.2. *Ophrys umbilicata* subsp. *bucephala*

Subsp. *bucephala* flowers from late March to late April and mainly grows in exposed olive groves and garrigue as well as on south-facing roadside slopes.

The distribution of this subspecies encompasses Lesbos, Chios, Samos and westernmost Anatolia. On Lesbos it mainly occurs in the south, around the cities of Agios Isidores, Plomari and Megalochori. On Samos it has only been reported from the low hills around Moni Agia Triada. [Aeg, Tur; Ana].

DESCRIPTION. *Ophrys umbilicata* Desf. subsp. *bucephala* (Gölz & H. R. Reinhard) B. Biel
Syn. *O. bucephala* Gölz & H. R. Reinhard.

Plant compact (the uppermost foliage leaf usually reaching past the node of the lowermost flower), 10-20 cm tall with 2-8 flowers in a dense spike. **Sepals** green (to white), occasionally suffused with violet, 12-17.5 × 6-9 mm. **Petals** 5-8 × 2.5-5 mm. **Lip** 12-15.5 × 15-19 mm; side lobes with straight apex. FIG. 50.

REFERENCES: Baumann & Künkele (1982, 1988), Buttler (1986), E. Danesch & O. Danesch (1969), Davies et al. (1983), Delforge (1994, 2001, 2005), Faurholdt (2003a), Nelson (1962), Paulus & Gack (1990a, 1990b), Pedersen & Faurholdt (1997a), Saliaris (2002), Sundermann (1980), Völh (1984).

FIG. 50. *Ophrys umbilicata* subsp. *bucephala* (**A.** Greece, Lesbos, Plomari, 31st March 1996; **B.** Greece, Lesbos, Plomari, 27th March 2002). Photos by N. Faurholdt (**A**), A. Gøthgen (**B**).

MAP 13. The European range of *Ophrys scolopax*. Six subspecies are recognized in Europe, where the highest diversity is found in Italy and in Greece (four subspecies each).

11. *Ophrys scolopax*

Surprisingly, this species is almost absent from mainland Italy (which is otherwise rich in bee orchids), but it occurs commonly and with a great variety of forms both to the east and west of this region. *Ophrys scolopax* resembles – and may be confused with – the widespread *O. fuciflora* and the pronouncedly eastern *O. umbilicata*. It differs from the former by its deeply three-lobed lip, the side lobes of which are converted into obliquely conical to horn-like bulges. It is distinguished from *O. umbilicata* by the orientation of the dorsal sepal which does not form a roof over the column in *O. scolopax*. It is involved in partly stabilised hybrid complexes with *O. fuciflora* (see H2. *O.* ×*vicina*) and *O. argolica* (see H3. *O.* ×*delphinensis*), respectively, which can render identification difficult locally. *Ophrys scolopax* is pollinated by various anthophorid bees of the genera *Eucera*, *Stilbeucera* and *Tetralonia*.

    Ophrys scolopax is very variable, and the morphological variation mainly falls between six groups of populations. These groups are also distinguished with regard to distribution and/or flowering time. Additionally, there appears to be a certain degree of pollinator specificity (an area in need of further study). Against this background, we recognise the six groups of populations as separate subspecies.

HABITAT. Dry as well as moist soil in full sunlight to light shade, from sea level to 2000 m altitude. Typical habitats include roadside slopes, garrigue, open pine and oak woods, grassland and olive groves that are run without application of fertiliser, pesticides and mechanical treatment of the soil.
FLOWERING TIME. From March to June, in most places with a distinct peak in April.
DISTRIBUTION. Throughout most of the Mediterranean and further east to the Crimea, Iran and the Caucasus. It seems absent from Cyprus and some other islands as well as from the Mediterranean coast of the Levant. In the northeast, it reaches Hungary and Rumania. [Aeg, Alb, Bal, Bul, Cor, Cre, Fra, Gre, Hun, Ita, Por, Rum, Sar, Sic, Spa, Tur, Ukr, Yug; Ana, Mor, Tun] MAP 13.
DESCRIPTION. *Ophrys scolopax* Cav.
Synonyms are listed under the subspecies.

Plant slender, 10-50(-90) cm tall with 2-15(-21) flowers in a lax spike.
Sepals violet to white or green, (ob)ovate to elliptic (less frequently lanceolate-oblong), 7-16 × 3-10 mm; dorsal sepal flat to shallowly boat-shaped, more or less incurved, from the base reflexed. **Petals** violet to rose-coloured or green, triangular to triangular-lanceolate or linear (sometimes auriculate) with reflexed margins, 1.5-8 × 0.8-4 mm, shaggy, spreading to slightly recurved. **Lip** with (reddish) brown to dark brown ground colour, straight with strongly recurved sides, deeply three-lobed close to the base, 6-16 × 6-20(-30) mm, brownish shaggy on the outer side of the side lobes and velvety along the margin of the mid-lobe (otherwise nearly glabrous); side lobes converted into obliquely conical to horn-like bulges; mid-lobe of varying length in relation to the side lobes, obtuse to rounded, provided with a broad and conspicuous, erect (to porrect), rectangular, rhomboid or (ob)triangular, often dentate appendage; **mirror** distinct, consisting of an H- or X-shaped to considerably more complicated (rarely simpler) figure, the basal arms of which are connected to the base of the lip, dull greyish blue to violet (rarely reddish brown) with an (often broad) cream border. **Column** acute to obtuse, not tapering towards the base (in side view); stigmatic cavity at least as wide as long and approximately twice as wide as the anther, with dark (rarely pale), lateral, eye-like knobs at base. FIGS 51-56.

Key to the subspecies

1. Side lobes of the lip horn-like, long-acuminate,
    (4-)6-12(-20) mm long . . . . 11.4. **subsp. *cornuta***
1. Side lobes of the lip obliquely conical (rarely horn-like),
    rounded to obtuse (less frequently acute), 1-6(-10) mm long . . . .
    2

2. Lip 13-16 mm long; appendage 2.5-5 mm long . . . .
    . . . . 11.5. **subsp. *heldreichii***
2. Lip 6-13(-15) mm long; appendage 1.5-2(-2.5) mm long 3

3. Sepals green (rarely white). Petals triangular to
    triangular-lanceolate 4
3. Sepals violet to white (rarely green, and in that case the petals
    are nearly always linear) 5
4. Plant 10-35 cm tall. Petals approximately as long as wide.
    Mirror covering almost the entire mid-lobe of the lip . . . .
    . . . . 11.6. **subsp. *rhodia***
4. Plant (10-)40-65 cm tall. **Petals** longer than wide. Mirror
    covering the basal half (rarely more) of the lip . . . .
    . . . . 11.3. **subsp. *conradiae***

5. Petals triangular to triangular-lanceolate. Mid-lobe of the lip
    not markedly narrowed at base . . . . 11.1. **subsp. *scolopax***
5. Petals more or less linear. Mid-lobe of the lip often markedly
    narrowed at base . . . . 11.2. **subsp. *apiformis***

11.1. *Ophrys scolopax* subsp. *scolopax*

In most places, this subspecies is easily recognised, although gradual
transitions can be seen to subsp. *heldreichii* on Rhodes, Paros and Antiparos,
to subsp. *apiformis* in the Iberian Peninsula and to subsp. *cornuta* in a number
of areas. Subsp. *scolopax* varies strongly in flower size, sepal colour, mirror
shape and the size of the lip side lobes. To us, this variation seems casual,
except that a group of Aegean populations deserves recognition as a separate
variety (see below). Flowering takes place from March to June with a peak
in April. Frequently, both early- and late-flowering individuals occur together.

The distribution encompasses most of the Mediterranean and continues
through Anatolia to Iran and the Caucasus. However, it should be noted
that this subspecies is sparse in mainland Italy and apparently absent from,
for example, Mallorca, Sardinia, Sicily, Crete and Cyprus. [Aeg, Alb, Bal,
Cor, Fra, Gre, Ita, Por, Spa, Ukr, Yug; Ana, Tun].

DESCRIPTION. *Ophrys scolopax* Cav. subsp. *scolopax*
Syn. [var. *scolopax*]: *O. bremifera* Steven; *O. oestrifera* M. Bieb. subsp. *bremifera*

(Steven) K. Richt.; *O. ceto* P. Devillers et al.; *O. corbariensis* J. Samuel &
J.-M. Lewin; *O. santonica* J. M. Mathé & Melki; *O. scolopax* Cav. subsp.
santonica (J. M. Mathé & Melki) R. Engel & P. Quentin; *O. fuciflora* (F. W.
Schmidt) Moench subsp. *scolopax* (Cav.) H. Sund.
Syn. [var. *minutula* (Gölz & H. R. Reinhard) H. A. Pedersen & N.
Faurholdt]: *O. minutula* Gölz & H. R. Reinhard.

Plant 10-50(-90) cm tall with 2-15(-21) flowers. **Sepals** violet to white
(very rarely green), 7-16 × 3-10 mm. **Petals** triangular to triangular-
lanceolate, sometimes auriculate, 2.5-8 × 1.2-4 mm, longer than wide. **Lip**
6-13(-15) × 7-16(-19) mm; side lobes obliquely conical (very rarely horn-
like), rounded to obtuse (rarely acute), 1-6(-10) mm long; mid-lobe not
markedly narrowed at base; appendage 1.5-2(-2.5) mm long; **mirror**
covering almost the entire mid-lobe (rarely less). FIG. 51.

FIG. 51A

FIG. 51. *Ophrys scolopax*
subsp. *scolopax*. **A-C.** var.
scolopax; **D.** var. *minutula*
(**A.** Spain, Galicia, Monte do
Cido, 27th April 2001; **B.**
Greece, Naxos, Alyko, 17th
April 2000; **C.** France, Provence,
Drap, 5th May 1990; **D.** Greece,
Chios, Elinda, 14th April 2005).
Photos by H. Æ. Pedersen (**A,
C-D**), N. Faurholdt (**B**).

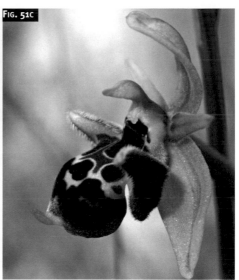

Var. *scolopax* is mainly characterised by its not distinctly secund spike, by its spreading to slightly reflexed sepals and by its lip which is at least 10 mm long. **Plant** 10-50(-90) cm tall with 2-15(-21) flowers in a not distinctly secund spike. **Sepals** spreading to slightly reflexed. **Petals** often auriculate. **Lip** 10-13(-15) × 11-16(-19) mm; side lobes 2-7(-10) mm long. [FIG. 51A-c]. This variety occurs throughout the geographic range of the subspecies. [Aeg, Alb, Bal, Cor, Fra, Gre, Ita, Por, Spa, Ukr, Yug; Ana, Tun].

Var. *minutula* is mainly characterised by its usually secund spike, by its reflexed sepals and by its lip which is up to 9 mm long. **Plant** 10-35 cm tall with 2-10 flowers in a secund spike. **Sepals** reflexed slightly to strongly. **Petals** rarely auriculate. **Lip** 6-9 × 7-10 mm; side lobes 1-3 mm long. [FIG. 51D]. The geographic range encompasses westernmost Anatolia and some of the eastern islands in the Aegean Sea (in Greece so far known from Lesbos, Chios and Samos). [Aeg; Ana].

129

11.2. *Ophrys scolopax* subsp. *apiformis*

Subsp. *apiformis* is considerably more variable than usually indicated in the literature, and gradual transitions to subsp. *scolopax* are frequent in Andalusia. In Portugal, subsp. *apiformis* is generally less well-defined than in other parts of its range (for example, the mid-lobe is less markedly narrowed at base). Even in Tunisia, however, similar plants usually constitute a minor part of otherwise typical populations of subsp. *apiformis*. Flowering takes place from March to June, with a peak in April in Europe.

The distribution is western Mediterranean and mainly encompasses Portugal, Andalusia and Mediterranean North Africa. Isolated occurrences are found in Sicily (Pantelleria) and Sardinia (Laconi) and possibly in the Balearic Islands (Ibiza). The taxonomic affinities of small-flowered populations of O. *scolopax* in Provence and the Pyrenees need further clarification – these populations, too, might be referable to subsp. *apiformis*. [Por, Sar, Sic, Spa; Mor, Tun].

DESCRIPTION. *Ophrys scolopax* Cav. subsp. *apiformis* (Desf.) Maire & Weiller Syn. O. *fuciflora* (F. W. Schmidt) Moench subsp. *apiformis* (Desf.) H. Sund.; O. *picta* Link; O. *sphegifera* Willd.

Plant 10-40 cm tall with 2-12(-14) flowers. **Sepals** rose-coloured to white (less frequently green), 7-11 × 4-6.5 mm. **Petals** more or less linear, rarely slightly auriculate, 2.5-4.5 × 0.8-1.5(-2) mm, much longer than wide. Lip 6-10 × 6-11 mm; side lobes obliquely conical, obtuse, 1-4 mm long; mid-lobe often markedly narrowed at base; appendage 1.5-2 mm long; mirror covering almost the entire mid-lobe. FIG. 52.

FIG. 52A

FIG. 52. *Ophrys scolopax* subsp. *apiformis* (**A.** Spain, Andalusia, Zahara, 18th April 2001; **B.** Portugal, the Algarve, Silves, 4th April 1999). Photos by H. Æ. Pedersen (**A**), N. Faurholdt (**B**).

FIG. 52B

FIG. 53. *Ophrys scolopax* subsp. *conradiae* (**A.** Italy, Sardinia, 30th May 1999; **B.** Italy, Sardinia, 9th June 2002). Photos by C. Giotta.

11.3. *Ophrys scolopax* subsp. *conradiae*

This subspecies is easily recognised and varies only a little. The flowering season is remarkably late, from early May to mid-June.

Subsp. *conradiae* is probably endemic to Sardinia and Corsica, although reports from Tunisia should be checked. [Cor, Sar].

DESCRIPTION. *Ophrys scolopax* Cav. subsp. *conradiae* (Melki & Deschâtres) H. Baumann et al.
Syn. *O. conradiae* Melki & Deschâtres; ?*O. scolopax* Cav. subsp. *sardoa* H. Baumann et al.

Plant (10-)40-65 cm tall with 3-10 flowers. **Sepals** green (rarely white), 9-12 × 5-6 mm. **Petals** more or less triangular, often auriculate, 3-4 × 2-3 mm, longer than wide. **Lip** 10-12 × 12-13 mm; side lobes obliquely conical, obtuse to rounded, 2-4 mm long; mid-lobe not markedly narrowed at base; appendage 1.5-2(-2.5) mm long; **mirror** covering the basal half (rarely more) of the lip. FIG. 53.

11.4. *Ophrys scolopax* subsp. *cornuta*

Subsp. *cornuta* varies considerably with regard to flower size and the size and shape of the side lobes of the lip. On a number of isles in the Aegean Sea, small-flowered plants with lip side lobes of varying size can be encountered.

FIG. 54. *Ophrys scolopax* subsp. *cornuta* (**A.** Greece, Pindhos, Agia Triada, 10th June 1998; **B.** Greece, Rhodes, Plimmiri, 16th April 1996; **C.** Greece, Chios, Elinda, 14th April 2005; **D.** Greece, Rhodes, Katavia, 26th March 2002). Photos by H. Æ. Pedersen (**A-C**), N. Faurholdt (**D**).

render identification difficult. The habitats are the same as indicated under the species, though almost consistently on calcareous ground. Flowering takes place from March to July; towards the end of this period, flowering specimens are exclusively found in mountainous areas in the northern part of the range.

The distribution is mainly eastern Mediterranean, stretching from the Caucasus across the Crimea and the Balkans to southern Hungary in the north, to Monte Gargano on the Italian east coast in the west, and to the Peloponnese, the Aegean Islands and Anatolia in the south. [Aeg, Bul, Gre, Hun, Ita, Rum, Tur, Ukr, Yug; Ana].

DESCRIPTION. *Ophrys scolopax* Cav. subsp. *cornuta* (Steven) E. G. Camus Syn. *O. bicornis* Sadler; *O. cerastes* P. Devillers & J. Devillers-Terschuren; *O. fuciflora* (F. W. Schmidt) Moench subsp. *cornuta* (Steven) H. Sund.; *O. cornuta* Steven; *O. cornutula* Paulus; *O. crassicornis* (Renz) J. Devillers-Terschuren & P. Devillers; *O. leptomera* P. Delforge; *O. oestrifera* M.-Bieb. s.s.; *O. scolopax* Cav. subsp. *oestrifera* (M.-Bieb.) Soó; *O. rhodostephane* P. Devillers & J. Devillers-Terschuren; *O. sepioides* P. Devillers & J. Devillers-Terschuren.

Plant 20-50 cm tall with 3-15 flowers. **Sepals** violet to rose-coloured (rarely white), 9-14 × 4-7 mm. **Petals** more or less triangular, usually auriculate, 1.5-5 × 1.5-3 mm, usually slightly longer than wide. **Lip** (8-)10-14 × 15-20(-30) mm; side lobes horn-like, long-acuminate (sometimes ending in almost thread-like tips), (4-)6-12(-20) mm long; mid-lobe not markedly narrowed at base; appendage 1.5-2(-2.5) mm long; **mirror** covering almost the entire mid-lobe (less frequently only its basal half). FIG. 54.

11.5. *Ophrys scolopax* subsp. *heldreichii*

In Crete, where no other representatives of *O. scolopax* are found, subsp. *heldreichii* is very distinct. Further north, on the other hand, it is more variable and often forms gradual transitions to subsp. *scolopax*. Variation is primarily seen in the flower size and in the size of the side lobes and appendage of the lip. Hybridisation with *O. fuciflora* subsp. *fuciflora* contributes to the confusion, particularly in a number of Aegean Islands (see H2. *O. ×vicina*).

Flowering takes place from March to early May. In Crete, where this subspecies is abundant, the flowering peaks in the first half of April.

Subsp. *heldreichii* is endemic to Greece. So far, it has been recorded from the mainland and from the following Aegean Islands: Antiparos, Crete,

Euboea, Karpathos, Lipsoi, Naxos, Paros and Rhodes. Reports from Samos and Lesbos need verification. [Aeg, Cre, Gre].

DESCRIPTION. *Ophrys scolopax* Cav. subsp. *heldreichii* (Schltr.) E. Nelson Syn. *O. heldreichii* Schltr. s.s.; *O. heldreichii* Schltr. var. *schlechterana* (Soó) Soó.

Plant 15-45 cm tall with 2-10 flowers. **Sepals** violet to rose-coloured, 12-16 × 6-10 mm. **Petals** triangular to triangular-lanceolate, rarely distinctly auriculate, 3-6 × 2-3.5 mm, longer than wide. **Lip** 13-16 × 15-19 mm; side lobes obliquely conical (to horn-like), obtuse to acute, 3-6(-10) mm long; mid-lobe not markedly narrowed at base; appendage 2.5-5 mm long; **mirror** covering almost the entire mid-lobe. FIG. 55.

11.6. *Ophrys scolopax* subsp. *rhodia*

Subsp. *rhodia* is easily recognised and does not vary much. This subspecies grows fully exposed in dry to moist calcareous soil – most often in grassland, grassy garrigue and fallow fields, but also in waste ground (a fairly rare phenomenon in *Ophrys*). Flowering takes place from late March to early May, with a peak in mid-April.

Subsp. *rhodia* seems endemic to Rhodes (where it is rather common) and Karpathos (where it is very rare). It has also been reported from Chios, Cyprus and Israel, but the photographs seen by us show individuals of *O. umbilicata* subsp. *umbilicata* with small flowers and green sepals. [Aeg, Cre].

DESCRIPTION. *Ophrys scolopax* Cav. subsp. *rhodia* (H. Baumann & Künkele) H. A. Pedersen & N. Faurholdt
Syn. *O. rhodia* (H. Baumann & Künkele) P. Delforge; *O. umbilicata* Desf. subsp. *rhodia* H. Baumann & Künkele.

Plant 10-35 cm tall with 3-12 flowers. **Sepals** green, 10-13 × 4.5-7 mm. **Petals** more or less triangular, often auriculate, 3-4.2 × 3-4.2 mm, approximately as long as wide. **Lip** 7.5-12 × 7.5-12 mm; side lobes obliquely conical, obtuse, 2-5 mm long; mid-lobe not markedly narrowed at base; appendage 1.5-2 mm long; **mirror** covering almost the entire mid-lobe. FIG. 56.

REFERENCES: Baumann & Künkele (1982, 1988), Bournérias (1998), Buttler (1986), E. Danesch & O. Danesch (1969), Davies et al. (1983), Del Prete & Tosi (1988), Delforge (1994, 2001, 2005), Eberhardt (1995), Faurholdt (2003a), Kullenberg (1961, 1973), Nelson (1962), Paulus (2001b), Paulus & Gack (1981, 1986), Paulus et al. (1983), Rossi (2002), Saliaris (2002), Souche (2004), Sundermann (1980), Vöth (1984, 1987).

FIG. 56. *Ophrys scolopax* subsp. *rhodia* (**A.** Greece, Rhodes, Thermes Kalithea, 25th March 1989; **B.** Greece, Rhodes, Kamiros, 15th April 1996). Photos by A. Gøthgen (**A**), H. Æ. Pedersen (**B**).

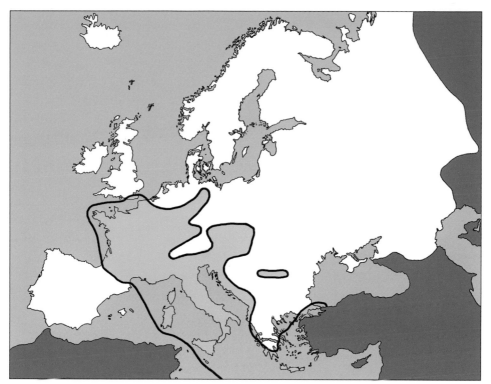

Map 14. The European range of *Ophrys fuciflora*. Ten subspecies are recognized in Europe, where the highest diversity is found in Italy (eight subspecies).

12. *Ophrys fuciflora*

Ophrys fuciflora is one of the most conspicuous bee orchids, partly due to the fact that the plant is usually tall and many-flowered, but particularly due to its colourful and fairly large flowers. This species resembles *O. scolopax*, but it is distinguished by the (usually) entire lip, the bulges of which are distinctly isolated from the margin. The so-called *O. philippei* Gren., *O. truncata* Dulac and *O. vetula* Risso are morphologically intermediate and probably consist of hybrids between the two species (but we have no personal experience of these French entities). Incidentally, *O. fuciflora* is regionally involved in partly stabilised hybrid complexes with *O. scolopax* (see H2. *O.* ×*vicina*) and *O. sphegodes* (see H4. *O.* ×*arachnitiformis*), respectively, which may render identification difficult.

    Ophrys fuciflora is highly variable, and the morphological variation mainly falls between 10 groups of populations. These are also different with regard to distribution, flowering season and/or habitat preferences. There seems to be a certain degree of pollinator specificity among the 10 groups of populations, but this is still insufficiently studied. Most remarkably, subsp.

chestermanii is the only *Ophrys* known to be pollinated by a cuckoo bumble bee, *Psithyrus vestalis* (family Apidae). Otherwise, O. *fuciflora* is mainly pollinated by various anthophorid bees of the genera *Eucera*, *Hetereucera*, *Tetralonia* and *Tetraloniella*. Additional pollinators are the megachilid bees *Chalicodoma ericetorum* and C. *pyrenaica*. Against this background we recognise the 10 entities as subspecies.

HABITAT. Dry as well as moist soil in full sunlight to light shade, from sea level to 1500 m altitude. Typical habitats include garrigue, roadside slopes, open pine and oak woods, grassland and old, pesticide-free olive groves.
FLOWERING TIME. In most places, flowering takes place from March to June with a distinct peak from mid-April to late May. Subsp. *elatior*, however, is very late-flowering, from June to September.
DISTRIBUTION. Central Europe, in the central and eastern Mediterranean and further east to Iraq. It appears to be absent from the Iberian Peninsula (however, reports from Spain should be checked) and the Balearic Islands as well as from the major part of the Balkan Peninsula. In North Africa, it is known from Libya only. [Aeg, Alb, Aus, Bel, Cor, Cre, Cze, Eng, Fra, Ger, Gre, Hol, Hun, Ita, Lux, Mal, Rum, Sar, Sic, Swi, Tur, Yug; Ana, Cyp, Isr] MAP 14.
DESCRIPTION. *Ophrys fuciflora* (F. W. Schmidt) Moench
Synonyms are listed under the subspecies.

Plant compact to slender, 10-80(-90) cm tall with 2-13(-21) flowers in a lax to dense spike. **Sepals** (purplish) violet to white or (pale) green, (ob)ovate to elliptic (less frequently oblanceolate-oblong), 9-19 × 3-11 mm; dorsal sepal boat-shaped to nearly flat, slightly (to strongly) incurved, from the base reflexed. **Petals** (purplish) violet to rose-coloured or (pale) green, (narrowly) triangular to triangular-oblong (often auriculate) with recurved margins, 1.5-9 × 1.5-5 mm, shaggy, spreading. **Lip** with (yellowish) brown to purplish black ground colour and sometimes a (very) broad, light brown to yellow margin, straight, flat or with more or less recurved sides, entire (to slightly three-lobed near the middle), 6-18 × 6-24 mm, velvety to shaggy along the edge (predominantly or exclusively towards the base), otherwise (sub)glabrous; bulges weakly developed to obliquely conical, distinctly isolated from the margin of the lip; front edge rounded to emarginate (rarely obtuse), provided with a broad and conspicuous, erect to porrect, rectangular, rhombic or (ob)triangular, often dentate appendage; **mirror** distinct, usually consisting of an H-shaped to considerably more complicated figure, the basal arms of which are distinctly connected to the base of the lip, less frequently marbled or consisting of more or less isolated spots and dashes, dull greyish blue to violet (rarely reddish brown), usually with a (sometimes very broad) cream border. **Column** acute (to obtuse), not tapering towards the base (in side view); stigmatic cavity approximately as wide as long and approximately twice as wide as the anther, with dark (rarely pale), lateral, eye-like knobs at base. FIGS 57–66.

Key to the subspecies

1. Mirror pronouncedly marbled, delimited by a broad, nearly unbranched, cream band .12.4. **subsp. *candica***
1. Mirror consisting of well-defined figures (rarely marbled); often provided with a pale border, but if this is broad and cream, it is nearly always distinctly branched 2

2. Sepals at least four times as long as the petals. Petals approximately as long as wide 3
2. Sepals not more than three times as long as the petals. Petals distinctly longer than wide 5

3. Lip trapeziform (its maximum width distinctly less than twice its width at base), at the centre with reddish to yellowish brown ground colour . . . . 12.10. **subsp. *biancae***
3. Lip flabellate (its maximum width at least twice its width at base), at the centre with dark reddish brown ground colour 4

4. Lip nearly straight, with dark reddish brown ground colour (occasionally, however, provided with a broad, light green to yellow margin); mirror covering more than one third of the lip . . . . 12.9. **subsp. *oxyrrhynchos***
4. Lip sigmoidly curved in longitudinal section, centrally with dark reddish brown ground colour, but always provided with a very broad, yellow margin; mirror covering less than one third of the lip . . . . .12.8. **subsp. *lacaitae***

5. Mirror reduced and fragmented, consisting of more or less isolated spots and dashes (rarely of more complex, but still isolated figures) . . . . .12.5. **subsp. *andria***
5. Mirror coherent, simple to complicated, often incorporating a larger, H-shaped figure 6

6. Mirror small (approximately half as wide as the lip), simple, more or less H-shaped 7
6. Mirror large (approximately two thirds as wide as the lip), usually complicated 8

7. Sepals green to white (rarely violet). Lip circular-quadrate, with medium brown ground colour . . . . 12.6. **subsp. *parvimaculata***

7. Sepals white to violet. Lip trapezoid (to broadly obovate), with dark brown to purplish black ground colour . . . . 12.7. **subsp. *chestermanii***

8. Petals (4-)6-9 mm long. Lip with sigmoidly recurved sides; appendage straight (rarely upcurved). . . . . 12.3. **subsp. *apulica***

8. Petals 3-6 mm long. Lip with spreading or more or less (but not sigmoidly) recurved sides; appendage upcurved 9

9. Foliage leaves still fresh at the peak of flowering. Sepals 10-16 mm long . . . . 12.1. **subsp. *fuciflora***

9. Foliage leaves withered at the peak of flowering. Sepals 8-10 mm long . . . . 12.2. **subsp. *elatior***

12.1. *Ophrys fuciflora* subsp. *fuciflora*

This subspecies, when delimited as broadly as here, is by far the most widespread and variable one. Other authors have in recent years proposed a pronounced splitting of this entity – even at species level. For example, a series of narrowly defined "species" from Corsica, Sardinia and Sicily as well as from the central and southern parts of mainland Italy have been segregated on the basis of differently sized flowers and lip bulges and differently coloured sepals (sometimes supported by somewhat divergent flowering seasons). Very often, however, a remarkable variation in the critical characters can be observed in one and the same population. In the eastern part of the range, the picture is even more muddled, and we feel that the systematic validity of the narrowly delimited entities needs to be further documented. Flowering normally takes place from late March to early July, in the Mediterranean with a peak from mid-April to late May. In central and southern Italy, flowering populations can still be found in mid-July.

The distribution encompasses almost the entire range of the species. However, subsp. *fuciflora* is absent from, for example, Malta, Cyprus and a number of islands in the central part of the Aegean Sea. [Aeg, Alb, Aus, Bel, Cor, Cre, Cze, Eng, Fra, Ger, Gre, Hol, Hun, Ita, Lux, Rum, Sar, Sic, Swi, Tur, Yug; Ana, Isr].

DESCRIPTION. *Ophrys fuciflora* (F. W. Schmidt) Moench subsp. *fuciflora* Syn. *O. aegirtica* P. Delforge; ?*O. annae* J. Devillers-Terschuren & P. Devillers; ?*O. fuciflora* (F. W. Schmidt) Moench subsp. *annae* (J. Devillers-Terschuren & P. Devillers) R. Engel & P. Quentin; *O. brachyotes* Rchb.f.; *O. episcopalis* Poir.; *O. halia* Paulus; *O. helios* Kreutz; *O. holoserica* s.s. auct., non (Burm.f.) Greuter; *O. linearis* (Moggr.) P. Delforge et al.; *O. holoserica* (Burm.f.) Greuter var. *linearis* (Moggr.) Landwehr, comb. inval.; *O. fuciflora*

FIG. 57A

FIG. 57B

FIG. 57C

FIG. 57. *Ophrys fuciflora* subsp. *fuciflora* (**A.** Greece, Samos, Pyrgos, 9th April 1998; **B.** Greece, Rhodes, Kamiros, 15th April 1996; **C.** Italy, Toscana, Castagneto, 12th April 2001; **D.** Croatia, Istria, Draguc, 21st May 2004; **E.** Greece, Rhodes, Thermes Kalithea, 29th March 2002). Photos by N. Faurholdt.

FIG. 57D

FIG. 57E

(F. W. Schmidt) Moench subsp. *lorenae* De Martino & Centurione; *O. fuciflora* (F. W. Schmidt) Moench subsp. *maxima* (H. Fleischm.) Soó; *O. holoserica* (Burm.f.) Greuter var. *maxima* (H. Fleischm.) Landwehr, comb. inval.; *O. medea* P. Devillers & J. Devillers-Terschuren; *O. minoa* (C. & A. Alibertis) P. Delforge; *O. candica* Greuter et al. subsp. *minoa* C. & A. Alibertis; ?*O. serotina* Rolli ex Paulus; *O. untchjii* (M. Schulze) P. Delforge.

Plant 10-40(-50) cm tall with 2-10 flowers. Foliage leaves still fresh at the peak of flowering. **Sepals** white to dark violet (very rarely green), 10-16 × 5-10 mm, 2.5-3 times as long as the petals. **Petals** 3-6 × 2-4 mm. **Lip** with dark brown to reddish brown ground colour, nearly straight, quadrate-trapezoid (to broadly obovate), more or less flat (less frequently with somewhat recurved sides), 8-12 × 12-24 mm; appendage upcurved, less than half as long as the column; **mirror** covering half to nearly three fourths of the lip, H-shaped to (usually) complicated with a branched, cream border. FIG. 57.

FIG. 58. *Ophrys fuciflora* subsp. *elatior* (**A-B.** Germany, Baden-Württemberg, Steinenstadt, 5th July 1992). Photos by C. A. J. Kreutz.

12.2. *Ophrys fuciflora* subsp. *elatior*

Subsp. *elatior* resembles subsp. *fuciflora*, but it is fairly easy to distinguish by the features indicated in the key and by its significantly later flowering. The habitats include grassland as well as open scrubs and woods, usually on calcareous ground. This subspecies flowers later than any other bee orchid – from early July (or late June) to late August, with flowering individuals occasionally encountered in September.

The distribution seems to be central European, covering areas in and around the upper Rhine and Rhône valleys. However, late-flowering entities described from Italy (*O. gracilis*, *O. posidonia*) and Istria (*O. tetraloniae*) may also belong here. [Fra, Ger, Swi].

DESCRIPTION. *Ophrys fuciflora* (F. W. Schmidt) Moench subsp. *elatior* Gumprecht ex R. Engel & P. Quentin
Syn. *O. elatior* Paulus; *O. holoserica* (Burm.f.) Greuter subsp. *elatior* (Gumpr. ex R. Engel & P. Quentin) H. Baumann & Künkele; ?*O. gracilis* (Büel et al.) Englmaier; ?*O. holoserica* (Burm.f.) Greuter subsp. *gracilis* (Büel et al.) O. & E. Danesch; ?*O. posidonia* P. Delforge; ?*O. tetraloniae* W. P. Teschner.

Plant (12-)20-80(-90) cm tall with (2-)6-13(-21) flowers. Foliage leaves withered at the peak of flowering. **Sepals** white to dark violet, 8-10 × 3-5.5 mm, 2-3 times as long as the petals. **Petals** 3-4.5 × 1.5-4 mm. **Lip** with dark brown to reddish brown ground colour, nearly straight, quadrate-trapezoid to broadly obovate, with gradually recurved sides (rarely flat), 6-12 × 6-14 mm; appendage upcurved, less than half as long as the column; **mirror** covering half to three fourths of the lip, H-shaped to (usually) a little more complicated with a branched, cream border. FIG. 58.

12.3. *Ophrys fuciflora* subsp. *apulica*

Subsp. *apulica* is usually easily recognised, but it does exhibit some variation. Individuals with a slightly three-lobed lip may resemble *O. scolopax* subsp. *heldreichii* (which, however, does not occur in the same area), and, particularly in late-flowering populations, individuals somewhat reminiscent of *O. scolopax* subsp. *scolopax* are often encountered. Flowering takes place from late March to June, with a peak from mid-April to early May.

This subspecies appears endemic to southern Italy, from Abruzzo southwards, although *O. dinarica* and *O. pharia*, originally described from Dalmatia, might belong in its synonymy. *Ophrys fuciflora* subsp. *apulica* is abundant in southern Puglia and on Monte Gargano. In Sicily, it is very rare and restricted to coastal areas in the south. Reports from some of the

Aegean Islands all seem erroneous and should be referred to *O.* ×*vicina* nm. *"calypsus"*. [Ita, Sic].

DESCRIPTION. *Ophrys fuciflora* (F. W. Schmidt) Moench subsp. *apulica* O. & E. Danesch
Syn. *O. apulica* (O. & E. Danesch) O. & E. Danesch; *O. holoserica* (Burm.f.) Greuter subsp. *apulica* (O. & E. Danesch) Buttler; ?*O. dinarica* R. Kranjcev & P. Delforge; ?*O. pharia* P. Devillers & J. Devillers-Terschuren.

Plant 15-35(-60) cm tall with 3-10(-14) flowers. Foliage leaves still fresh at the peak of flowering. **Sepals** rose-coloured to violet, (12-)14-19 × 7-10 mm, 2-2.5(-3) times as long as the petals. **Petals** (4-)6-9 × 3-5 mm. **Lip** with light brown to dark reddish brown ground colour, nearly straight, quadrate-trapezoid (to broadly obovate) with sigmoidly recurved sides, 14-18 × 17-22 mm; appendage porrect (rarely upcurved), usually about half as long as the column (occasionally a little shorter); **mirror** covering approximately half of the lip, H-shaped to more complicated with a branched, cream border. FIG. 59.

FIG. 59. *Ophrys fuciflora* subsp. *apulica* (**A.** Italy, Puglia, Martina Franca, 16th May 1995; **B.** Italy, Puglia, Frigole, 20th April 2002). Photos by N. Faurholdt.

FIG. 60. *Ophrys fuciflora* subsp. *candica* (**A**. Greece, Rhodes, Petaloudes, 20th April 1996; **B**. Italy, Puglia, San Cataldo, 21st April 2002). Photos by H. Æ. Pedersen (**A**), N. Faurholdt (**B**).

12.4. *Ophrys fuciflora* subsp. *candica*

Genetically pure subsp. *candica* is easily recognised and varies very little. Apparent hybridisation with other bee orchids, however, may locally render identification difficult and has even led to description of dubious species. For instance, we believe that *O. tardans* O. & E. Danesch from southern Puglia consists of fortuitous hybrids between *O. fuciflora* subsp. *candica* and *O. tenthredinifera*. Similarly, we believe that Sicilian populations referred to as *O. calliantha* consist partly of pure subsp. *candica*, partly of hybrids between the latter and subsp. *oxyrrhynchos*. The hybrids can be recognised from their often large, porrect lip appendage and their small petals. In Sicilian plants, the mirror is usually characteristic of subsp. *candica* (i.e. marbled, delimited by a broad, cream band and with only a small basal field). However, individuals are fairly commonly seen that have a mirror more reminiscent of subsp. *fuciflora*, often incorporating a central, eye-like spot. The peak flowering of subsp. *candica* and the hybrid is later than in subsp. *oxyrrhynchos*. Flowering takes place in April and May, with a peak from late April to mid-May.

The distribution is fragmented and encompasses several areas in the central and eastern Mediterranean. Confirmed finds are known from Sicily, Puglia, Basilicata, the southern Peloponnese, Kithira, Crete, Rhodes and southwestern Anatolia. Reports from Samos should be checked, and a more widespread occurrence in Greece does not seem unlikely. [Aeg, Cre, Ita, Sic; Ana].

DESCRIPTION. *Ophrys fuciflora* (F. W. Schmidt) Moench subsp. *candica* Soó [non E. Nelson, nom. inval.]

Syn. ?*O. calliantha* Bartolo & Pulvirenti; *O. candica* (Soó) H. Baumann & Künkele [non Greuter et al., nom. illeg.]; *O. fuciflora* (F. W. Schmidt) Moench var. *candica* (E. Nelson) H. Sund., comb. inval.; *O. holoserica* (Burm.f.) Greuter subsp. *candica* (Soó) Renz & Taubenheim; *O. holoserica* (Burm.f.) Greuter var. *candica* (E. Nelson) H. Sund., comb. inval. ?*O. lacaena* P. Delforge.

Plant 15-45 cm tall with 2-7 flowers. Foliage leaves still fresh at the peak of flowering. **Sepals** rose-coloured to white (less frequently violet), 11-15 × 4-8 mm, 3-5(-6) times as long as the petals. **Petals** 2-4.5 × 1.5-3 mm. **Lip** with reddish brown to dark brown ground colour, nearly straight, circular-quadrate with somewhat recurved sides (less frequently flat), 9-14 × 12-16 mm; appendage upcurved, less than half as long as the column; **mirror** covering half to two thirds of the lip, pronouncedly marbled, delimited by a broad, (very nearly) unbranched, cream band. FIG. 60.

12.5. *Ophrys fuciflora* subsp. *andria*

In most cases this subspecies is easy to recognise, although problems may arise in connection with individuals in which markings of the mirror are diffusely connected to the base of the lip. It is notable that the populations

FIG. 61. *Ophrys fuciflora* subsp. *andria* (**A-B.** Greece, Naxos, Koronos, 18th April 2000). Photos by N. Faurholdt.

on Andros consist almost exclusively of individuals with green or greenish rose-coloured sepals, whereas apparently all plants on Naxos have rose-coloured to dark violet sepals. Flowering is rather early, from mid-March to mid-April.

Subsp. *andria* is endemic to the Cyclades in the Aegean Sea. So far, it has been recorded from Andros, Naxos, Tinos and Kimolos. [Gre].

DESCRIPTION. *Ophrys fuciflora* (F. W. Schmidt) Moench subsp. *andria* (P. Delforge) N. Faurholdt
Syn. ?*O. aeoli* P. Delforge; *O. andria* P. Delforge s.s.; *O. andria* P. Delforge var. *halkionis* G. & H. Kretzschmar; *O. thesei* P. Delforge.

Plant 15-40(-50) cm tall with 2-10 flowers. Foliage leaves still fresh at the peak of flowering. **Sepals** green to rose-coloured or dark violet, 10-16 × 5-11 mm, 2-3 times as long as the petals. **Petals** (3-)4-7.5 × (2-)2.5-4 mm. **Lip** with dark brown to reddish brown ground colour, nearly straight, circular-quadrate to trapezoid (rarely broadly obovate), more or less flat, 12-16 × 14-20 mm; appendage upcurved, less than half as long as the column; **mirror** reduced and fragmented, consisting of more or less isolated spots and dashes (rarely of more complex, but still isolated, figures) with cream borders. FIG. 61.

12.6. *Ophrys fuciflora* subsp. *parvimaculata*

This subspecies is easily recognised and exhibits little variation. Typical habitats include open woods and scrubs of *Quercus pubescens* (downy oak) and *Ostrya carpinifolia* (hop hornbeam), from sea level to 600 m altitude. Less frequently, subsp. *parvimaculata* grows in grassland, often among shrubs of *Paliurus spina-christi* (Christ's thorn). Flowering takes place in April and May, with a peak from mid-April to early May.

Subsp. *parvimaculata* is endemic to southern Italy, where it is found in Puglia (from Monte Gargano to Lecce) and Basilicata (the province of Matera). On Monte Gargano it is mainly found in the northern part, between lakes Lesiana and Varano [Ita].

DESCRIPTION. *Ophrys fuciflora* (F. W. Schmidt) Moench subsp. *parvimaculata* O. & E. Danesch
Syn. *O. parvimaculata* (O. & E. Danesch) Paulus & Gack; *O. holoserica* (Burm.f.) Greuter subsp. *parvimaculata* (O. & E. Danesch) O. & E. Danesch.

Plant 10-35 cm tall with 2-7 flowers. Foliage leaves still fresh at the peak of flowering. **Sepals** green to white (rarely violet), 12-15.5 × 6.5-8.5 mm, 2-3 times as long as the petals. **Petals** 4-7 × 2-4 mm. **Lip** with medium brown ground colour, nearly straight, circular-quadrate, flat (less frequently with somewhat recurved sides), 10-14 × 15-19 mm; appendage upcurved, less than half as long as the column; **mirror** covering one sixth to one fourth of the lip, more or less H-shaped with a whitish to cream border. FIG. 62.

12.7. *Ophrys fuciflora* subsp. *chestermanii*

Subsp. *chestermanii* exhibits little variation, and only the occurrence of hybrids may render identification difficult. Certain authors consider the so-called *O.* ×*normanii* J. J. Wood a distinct species, but, in accordance with J. J. Wood, we interpret this taxon as fortuitous hybrids between *O. fuciflora* subsp. *chestermanii* and *O. tenthredinifera*. Subsp. *chestermanii* normally grows in light shade on moist, calcareous, often stony ground. It is typically found on mossy slopes in open scrubs of *Quercus ilex* (holm oak). Flowering takes place from late March to mid-May, with a peak in the second half of April and early May.

FIG. 62

FIG. 62. *Ophrys fuciflora* subsp. *parvimaculata* (Italy, Puglia, Monte Gargano, 12th April 1995). Photo by H. Æ. Pedersen.

Fig. 63. *Ophrys fuciflora* subsp. *chestermanii* (Italy, Sardinia, Ponte su Crabiolu, 20th April 1997).
Photo by H. Æ. Pedersen.

This subspecies is endemic to Sardinia, where it is known from the vicinity of Iglésias in the southwest and from the area between Baunei and Muravera in the east. [Sar].

DESCRIPTION. *Ophrys fuciflora* (F. W. Schmidt) Moench subsp. *chestermanii* (J. J. Wood) Blatt & Wirth
Syn. *O. chestermanii* (J. J. Wood) Gölz & H. R. Reinhard; *O. holoserica* (Burm.f.) Greuter subsp. *chestermanii* J. J. Wood.

Plant 10-30 cm tall with 2-5(-7) flowers. Foliage leaves still fresh at the peak of flowering. **Sepals** white to violet, 10-18 × 4-8 mm, 2.5-3 times as long as the petals. **Petals** 4-6.5 × 1.5-3 mm. **Lip** with dark brown to purplish black ground colour, nearly straight, trapeziform (to broadly obovate), more or less flat, 12-18 × 15-23 mm; appendage upcurved, less than half as long as the column; **mirror** covering one sixth to one fourth of the lip, more or less H-shaped (occasionally with a narrow, white border). FIG. 63.

12.8. *Ophrys fuciflora* subsp. *lacaitae*

This pretty and characteristic subspecies can only be confused with forms of (the usually earlier flowering) subsp. *oxyrrhynchos* with yellow lip margin — but see the key. Flowering takes place from late April to early June.

Apart from a strong population on the Croatian island of Vis and an old record from Malta, this subspecies is only known from southern Italy, including Sicily and the mainland provinces of Potenza, Salerno, Isernia, Latina, Molise and Foggia (Monte Gargano). [Ita, Mal, Sic, Yug].

FIG. 64A

FIG. 64. *Ophrys fuciflora* subsp. *lacaitae* (**A**. Italy, Molise, Vandra, 3rd June 2005; **B**. Italy, Sicily, Cassibile, 22nd April 1999). Photos by N. Faurholdt (**A**), H. Æ. Pedersen (**B**).

Fig. 64B

DESCRIPTION. *Ophrys fuciflora* (F. W. Schmidt) Moench subsp. *lacaitae* (Lojac.) H. Sund.
Syn. *O. lacaitae* Lojac.; *O. holoserica* (Burm.f.) Greuter subsp. *lacaitae* (Lojac.) W. Rossi; *O. oxyrrhynchos* Tod. subsp. *lacaitae* (Lojac.) Del Prete; *O. holoserica* (Burm.f.) Greuter subsp. *oxyrrhynchos* (Tod.) H. Sund. var. *lutea* (Tineo) Landwehr, comb. inval.

Plant 10–30 cm tall with 2–11 flowers. Foliage leaves still fresh at the peak of flowering. **Sepals** green (to white), 11–14 × 5–6.5 mm, (4.5–)5–8 times as long as the petals. **Petals** 1.5–3 × 1.5–3 mm. **Lip** with dark reddish brown ground colour on its central part, but always provided with a very broad

FIG. 65A

FIG. 65B

FIG. 65. *Ophrys fuciflora* subsp. *oxyrrhynchos* (**A,C.** Italy, Sicily, Ferla, 18th April 1999; **B.** Italy, Puglia, Martina Franca, 20th April 2002). Photos by H. Æ. Pedersen (**A,C**), N. Faurholdt (**B**).

FIG. 65C

yellow margin, sigmoidly curved in longitudinal section, flabellate with more or less spreading sides, 11–14 × 14–18.5 mm; appendage upcurved, more than half as long as the column; **mirror** covering one eighth to one fourth of the lip, H-shaped to slightly more complicated with a branched, white to cream border. FIG. 64.

12.9. *Ophrys fuciflora* subsp. *oxyrrhynchos*

Subsp. *oxyrrhynchos* is easy to identify, especially in Sicily, although forms with a broad, yellow lip margin may be confused with (the slightly earlier flowering) subsp. *biancae* and (the somewhat later flowering) subsp. *lacaitae,* but see the key. In Puglia the picture is complicated by frequent hybridisation with subsp. *apulica* (in our opinion, the so-called *O. celiensis* (O. & E. Danesch) P. Delforge is identical with the hybrid). Flowering takes place from March to May, with a peak in April.

Today, subsp. *oxyrrhynchos* is found in Sicily (where it is common) as well as in Calabria, Campania, Basilicata and southern Puglia. An additional record exists from Malta, but it is uncertain if the subspecies is extant on that island. [Ita, Mal, Sic].

DESCRIPTION. *Ophrys fuciflora* (F. W. Schmidt) Moench subsp. *oxyrrhynchos* (Tod.) Soó
Syn. *O. oxyrrhynchos* Tod. s.s.; *O. holoserica* (Burm.f.) Greuter subsp. *oxyrrhynchos* (Tod.) H. Sund.

Plant 10–30 cm tall with 3–9 flowers. Foliage leaves still fresh at the peak of flowering. **Sepals** green to white (occasionally with a rose-coloured tinge), 11–16 × 5–8 mm, 4–5 times as long as the petals. **Petals** 2.5–4 × 2–4 mm. **Lip** with dark reddish brown ground colour, occasionally provided with a broad light brown to yellow margin, nearly straight, flabellate, more or less flat, 9.5–12 × 14–18 mm; appendage upcurved, more than half as long as the column; **mirror** covering one third to half of the lip, H-shaped to (usually) more complicated with a branched, white to cream border. FIG. 65.

12.10. *Ophrys fuciflora* subsp. *biancae*

This subspecies is moderately variable and may be confused with forms of (the slightly later flowering) subsp. *oxyrrhynchos* – but see the key. Flowering takes place in March and April, with a peak in the middle of this period.

Subsp. *biancae* is endemic to Sicily. It is fairly rare, being known only from the limestone areas south of Etna (Siracusa, Ragusa, Catania) and from the vicinity of Palermo. [Sic].

FIG. 66. *Ophrys fuciflora* subsp. *biancae* (**A.** Italy, Sicily, Ferla, 18th April 1999; **B.** Italy, Sicily, Ferla, 17th April 1992). Photos by H. Æ. Pedersen (**A**), N. Faurholdt (**B**).

DESCRIPTION. *Ophrys fuciflora* (F. W. Schmidt) Moench subsp. *biancae* (Tod.) N. Faurholdt & H. A. Pedersen

Syn. *O. biancae* (Tod.) Macch.; *O. discors* Bianca, nom. illeg.

Plant 10-25 cm tall with 4-10 flowers. Foliage leaves still fresh at the peak of flowering. **Sepals** white to light rose-coloured or greenish, 10-14 × 5-7 mm, 4-5 times as long as the petals. **Petals** 2-3 × 1.5-2.5 mm. **Lip** with reddish brown to yellowish brown ground colour, sometimes provided with a broad yellow margin, nearly straight, trapeziform, flat or with somewhat recurved sides, 8-11 × 10-15 mm; appendage upcurved, approximately half as long as the column; **mirror** covering one fourth to one third of the lip, H-shaped (to slightly more complicated) with an often branched, cream border. FIG. 66.

REFERENCES: Bartolo & Pulvirenti (1997), Baumann & Künkele (1982, 1988), Blatt & Wirth (1990), Bournérias (1998), Buttler (1986), E. Danesch & O. Danesch (1969), Davies et al. (1983), Del Prete & Tosi (1988), Delforge (1994, 2001, 2005), Faurholdt (2003), Gölz & Reinhard (2001), Gulli et al. (2003), Hirth & Spaeth (1998), Kranjčev (2001), Kreutz (2001), Kullenberg (1961), Nelson (1962), Paulus (1988, 1996, 2000, 2002), Paulus & Gack (1986, 1995), Pedersen & Faurholdt (1997), Reinhard (1987), Rossi (2002), Souche (2004), Sundermann (1980), Vöth (1987), Wirth & Blatt (1988), Wood (1982).

13. *Ophrys argolica*

The flowers of *O. argolica* often resemble small faces, which the eye-like markings of the mirror endow with expressions of happiness, sadness or astonishment. The "eyes" may also look inquisitively staring, or they may seem equipped with glasses. In most of the eastern Mediterranean, this species is easily recognised (cf. the shape and hairiness of the lip combined with the characteristic mirror), but in certain parts of Greece it forms a partly stabilised hybrid complex with *O. scolopax* (see H3. *O.* ×*delphinensis*). In a number of places in Italy, *O. argolica* forms bewildering hybrid swarms with *O. fuciflora* (see the section The classication developed for this book p. 55).

The morphological variation mainly falls between six groups of populations. These groups are distributed in fairly small geographic areas and hardly overlap. Additionally, each of the six entities seem to depend on their own specific pollinators. *Ophrys argolica* is predominantly pollinated by anthophorid bees, and the above groups of populations seem to be adapted to different species of *Anthophora* – apart from one group that is pollinated

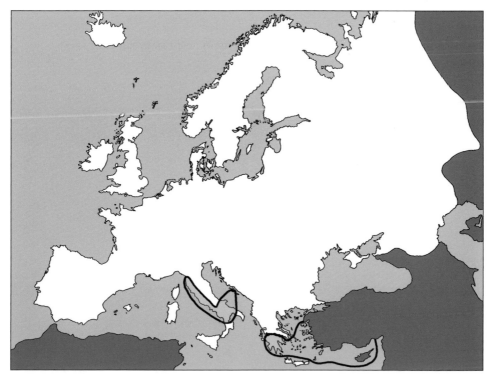

MAP 15. The European range of *Ophrys argolica*. Six subspecies are recognized in Europe, where the highest diversity is found in Greece (four subspecies).

by the andrenid bee *Andrena curiosa*. With this background, we recognise the six entities as subspecies.

HABITAT. Calcareous, dry to moist (rarely wet) soil in full sunlight to light shade, from sea level to 1300 m altitude. Typical habitats include garrigue, open oak and pine woods, grassland, roadside slopes and old pesticide-free olive groves.

FLOWERING TIME. From March to May, with most of the subspecies peaking in April.

DISTRIBUTION. Central and eastern Mediterranean, from mainland Italy across Greece to Anatolia and Cyprus, though missing on many islands in the Aegean Sea. [Aeg, Cre, Gre, Ita, Yug; Ana, Cyp] MAP 15.

DESCRIPTION. *Ophrys argolica* H. Fleischm.

Synonyms are listed under the subspecies.

Plant compact to slender, (7.5-)10-65 cm tall with (1-)2-8(-12) flowers in a dense to relatively lax spike. **Sepals** white to violet, less frequently (pale) green, narrowly (ovate-)elliptic, 10-18 × 5-10 mm; dorsal sepal (nearly) flat, straight to slightly incurved, from the base reflexed. **Petals** pale rose-coloured to (purplish) violet, (triangular-)ovate to lanceolate or (elliptic-)oblong with (nearly) flat margins, (3-)4-11 × 2-5.5 mm, shaggy to velvety, spreading or reflexed. **Lip** with (reddish) brown to yellowish brown or olive-green ground colour, straight and nearly flat, entire to moderately

FIG. 67. *Ophrys argolica* subsp. *crabronifera* (**A-B.** Italy, Toscana, Monte Argentario, 9th April 2001). Photos by N. Faurholdt.

three-lobed close to the middle, 9–16 × 12–20 mm, in the basal part more or less shaggy of usually light hairs along the margin (otherwise nearly glabrous – or velvety along the margin); bulges rarely present and in that case only weakly developed, distinctly isolated from the margin of the lip (side lobes flat, if present); front edge obtuse to emarginate, provided with a short, more or less porrect, awl-shaped to triangular (rarely somewhat rectangular) point; **mirror** distinct (rarely obscure), consisting of a horseshoe- to butterfly-shaped figure or two rhomboid to drop-shaped spots, placed centrally and isolated from the base of the lip (rarely connected to the latter by thin lines), dull (rarely shining) grey to blue, bluish violet or brownish to reddish violet, often with a pale border. **Column** acute to obtuse, not tapering towards the base (in side view); stigmatic cavity at least as wide as long and approximately twice as wide as the anther, with dark or pale, lateral, eye-like knobs at base. FIGS 67–72.

Key to the subspecies

1. Stigmatic cavity 4–6 mm wide at base . . . . . 13.1. **subsp. *crabronifera***
1. Stigmatic cavity 2–4 mm wide at base 2

2. Sepals reflexed . . . . 13.2. **subsp. *biscutella***
2. Sepals spreading 3

3. Basal half of the lip shaggy of long white hairs along the margin . . . . 13.3. **subsp. *argolica***
3. Basal half of the lip shaggy of relatively short, greyish to (purplish) brown (rarely whitish) hairs along the margin 4

4. Lip more or less three-lobed, widest in its basal part or at the middle . . . . 13.5. **subsp. *lucis***
4. Lip entire, widest close to the apex 5

5. Central part of the lip deep reddish brown; terminal point 0.5–1 mm wide at base. Stigmatic cavity whitish with a reddish brown line across . . . . .13.4. **subsp. *aegaea***
5. Central part of the lip (yellowish) brown to olive-green; terminal point 1–2 mm wide at base. Stigmatic cavity brownish with pale centre . . . . 13.6. **subsp. *lesbis***

13.1. *Ophrys argolica* subsp. *crabronifera*

The flowers of this subspecies are very variable with regard to colour of the sepals, appearance of the mirror and conformation of the lip. Subsp. *crabronifera* may be confused with subsp. *biscutella*, and individuals are occasionally encountered that are strongly reminiscent of the latter. Flowering takes place from March to May, with a peak in mid-April.

Subsp. *crabronifera* is endemic to a broad area along the Tyrrhenian coast of mainland Italy. There are many populations between Livorno and Rome, but further south (to around Salerno) it is rare and mainly coast-bound. Reports from Corsica should probably be referred to *O.* ×*arachnitiformis*. [Ita].

DESCRIPTION. *Ophrys argolica* H. Fleischm. subsp. *crabronifera* (Mauri) N. Faurholdt
Syn. *O. crabronifera* Mauri s.s.; *O. exaltata* s.s. auct., non Ten.; *O. fuciflora* (F. W. Schmidt) Moench subsp. *exaltata* auct., non (Ten.) E. Nelson; *O. holoserica* (Burm.f.) Greuter subsp. *exaltata* auct., non (Ten.) Landwehr, comb. inval.; *O. pollinensis* J. Devillers-Terschuren & P. Devillers; *O. fuciflora* (F. W. Schmidt) Moench subsp. *pollinensis* E. Nelson, nom. inval.; *O. holoserica* (Burm.f.) Greuter subsp. *pollinensis* O. & E. Danesch [non (E. Nelson) Landwehr, comb. inval.].

Plant 20-65 cm tall with 3-7(-12) flowers in a lax spike. **Sepals** whitish to rose-coloured or dark violet, less frequently green, (10-)12-17 × 6-8.5 mm, spreading to more or less reflexed. **Petals** shaggy, (3-)4-10 × 2.5-4.5 mm. **Lip** with brown to brownish olive-green ground colour on its central part, entire, widest at or above the middle, 11-15 × 13-19 mm, in (nearly) its whole length shaggy of fairly long, whitish to yellowish brown hairs along the margin; terminal point triangular to oblong (1.5-3 mm wide at base); **mirror** bluish grey with or without a pale border, with or (usually) without connection to the base of the lip. Stigmatic cavity 4-6 mm wide at base, brown to olive-green. FIG. 67.

13.2. *Ophrys argolica* subsp. *biscutella*

Subsp. *biscutella* varies a great deal as far as the appearance of the mirror is concerned. In some (mainly southern and western) populations, rather complicated markings are dominant and frequently they include lines that connect the mirror to the base of the lip. Particularly in Campania, a most confusing multitude of forms can be observed in populations which are more or less intermediary between *O. argolica* subsp. *biscutella* and *O. fuciflora* subsp. *fuciflora*, and which may even be accompanied by *O. argolica* subsp. *crabronifera*. We think that the variation in the latter populations is due to massive hybridisation and back-crossing (see also the section The classification developed for this book p.55). Flowering takes place from late March to mid-May, with a peak in mid-April.

The geographic range of subsp. *biscutella* encompasses the southern part of mainland Italy and the Croatian island of Korcula. The main occurrences

FIG. 68A

FIG. 68B

FIG. 68. *Ophrys argolica* subsp. *biscutella* (**A-B.** Italy, Puglia, Monte Gargano, 12th April 1995). Photos by A. Gøthgen.

in Italy are found on Monte Gargano (Puglia), Monte Alburni (Campania) and Monte Pollino (Calabria). [Ita, Yug].

DESCRIPTION. *Ophrys argolica* H. Fleischm. subsp. *biscutella* (O. & E. Danesch) Kreutz
Syn. *O. biscutella* O. & E. Danesch; *O. crabronifera* Mauri subsp. *sundermannii* (Soó) Del Prete, comb. inval.; *O. holoserica* (Burm.f.) Greuter subsp. *exaltata* (Ten.) Landwehr var. *sundermannii* (Soó) Landwehr, comb. inval.

Plant 10-50(-60) cm tall with 2-8(-10) flowers in a relatively dense to lax spike. **Sepals** white to rose-coloured or dark violet (rarely greenish), 13-18 × 6-9 mm, reflexed. **Petals** shaggy, 6-11 × 2-4.5 mm. **Lip** with (reddish) brown ground colour on its central part, entire, widest above the middle, 12.5-15 × 15.5-20 mm, the basal half (or more) shaggy of fairly long whitish to light brown hairs along the margin; terminal point (narrowly) triangular to oblong (1-3 mm wide at base); **mirror** bluish grey with or without a pale border, with or without connection to the base of the lip. Stigmatic cavity 3-4 mm wide at base, brownish. FIG. 68.

13.3. *Ophrys argolica* subsp. *argolica*

Though quite variable, subsp. *argolica* can be easily identified from the dense white hairiness of the lip margin. Flowering takes place from March to May, with a peak in April.

This subspecies is endemic to southern Greece, where it predominantly occurs in the Peloponnese and on Kithira between the Peloponnese and Crete. A recent record from Bulgaria seems referable to *O. ferrum-equinum*. [Gre].

DESCRIPTION. *Ophrys argolica* H. Fleischm. subsp. *argolica*.

Plant 15-30(-50) cm tall with 2-8(-10) flowers in a relatively dense spike. **Sepals** rose-coloured to (reddish) violet, 10-15.5 × 5-8 mm, spreading. **Petals** shaggy, 5-9 × 2.5-4 mm. **Lip** with (reddish) brown ground colour on its central part, entire to shallowly three-lobed, widest above the middle, 9-15 × 12-16.5 mm, the basal half shaggy of long white hairs along the margin; terminal point narrowly to broadly triangular (1-2 mm wide at base); **mirror** bluish grey to bluish violet with a pale border, with or (usually) without connection to the base of the lip. Stigmatic cavity (2-)3-4 mm wide at base, brownish with pale centre. FIG. 69.

FIG. 69A

FIG. **69.** *Ophrys argolica* subsp. *argolica* (**A-B.** Greece, the Peloponnese, Githion, 20th April 1997). Photos by A. Gøthgen.

FIG. 69B

FIG. 70A

FIG. 70B

FIG. 70C

FIG. 70. *Ophrys argolica* subsp. *aegaea* (**A.** Greece, Karpathos, Katodio–Spoa, 20th March 2000; **B-C.** Greece, Karpathos, Volada–Katodio, 20th March 2000). Photos by C. A. J. Kreutz.

13.4. *Ophrys argolica* subsp. *aegaea*

Subsp. *aegaea* is easily recognised and exhibits only modest variation. It chiefly occurs in garrigue with asphodels as well as in grassy groves of olive trees or carobs, up to 700 m altitude. Flowering is early, taking place from early to late March.

Subsp. *aegaea* was originally considered endemic to Karpathos, but it has recently been found also on the neighbouring island of Kasos and on the Cycladean islands of Amorgos and Iraklia. On the other hand, we suspect that an individual reported from Cyprus in 2001 is an aberrant plant of O. *argolica* subsp. *elegans* (Renz) E. Nelson. [Cre, Gre].

DESCRIPTION. *Ophrys argolica* H. Fleischm. subsp. *aegaea* (Kalteisen & H. R. Reinhard) H. A. Pedersen & N. Faurholdt
Syn. O. *aegaea* Kalteisen & H. R. Reinhard s.s.

163

Plant (7–)10–20(–30) cm tall with (2–)3–5(–7) flowers in a dense spike. **Sepals** whitish to rose-coloured or light reddish violet, 12–16 × 6–8.5 mm, spreading. **Petals** shaggy, 7–10.5 × 2.5–4 mm. **Lip** with deep (reddish) brown ground colour on its central part, entire, widest above the middle, 11–14 × 16–20 mm, its basal half shaggy of fairly short, greyish to light brown hairs along the margin; terminal point triangular (0.5–1 mm wide at base); **mirror** (greyish) blue with or without a pale border, without connection to the base of the lip. Stigmatic cavity 3–4 mm wide at base, whitish with a reddish brown line across. FIG. 70.

13.5. *Ophrys argolica* subsp. *lucis*

Subsp. *lucis* is only occasionally variable and can be readily identified (except in Anatolia where it can be confused with some of its close relatives). In Greece, it is somewhat reminiscent of subsp. *lesbis* – but see the key. Subsp. *lucis* is particularly often encountered in grassy garrigue and in open groves and woods dominated by pines or cypresses. Flowering mainly takes place in March, but often lasts to the beginning of April.

The distribution encompasses Rhodes (mainly the central part) and the nearby islands of Tilos and Nisiros as well as southwestern Anatolia. [Aeg; Ana].

FIG. 71. *Ophrys argolica* subsp. *lucis* (**A**. Greece, Rhodes, April 1987; **B**. Greece, Rhodes, Isidoros, 17th April 1996). Photos by B. Olsen (**A**), H. Æ. Pedersen (**B**).

FIG. 72. *Ophrys argolica* subsp. *lesbis* (**A-B.** Greece, Lesbos, Antissa, 9th April 1996). Photos by E. Jensen.

DESCRIPTION. *Ophrys argolica* H. Fleischm. subsp. *lucis* (Kalteisen & H. R. Reinhard) H. A. Pedersen & N. Faurholdt
Syn. *O. lucis* (Kalteisen & H. R. Reinhard) Paulus; *O. aegaea* Kalteisen & H. R. Reinhard subsp. *lucis* Kalteisen & H. R. Reinhard.

Plant 10-20(-30) cm tall with 2-4(-5) flowers in a relatively dense spike. **Sepals** white to rose-coloured or dark violet (rarely green), 11-15 × 6-8.5 mm, spreading. **Petals** shaggy, 6.5-9.5 × 2.5-4 mm. **Lip** with brownish olive-green to reddish brown ground colour on its central part, more or less three-lobed (very rarely entire), widest at or below the middle, 11-13 × 14-19 mm, the basal half shaggy of fairly short, greyish to purplish brown hairs along the margin; terminal point triangular (1-2 mm wide at base); **mirror** bluish grey to brownish violet with or without a pale border, without connection to the base of the lip. Stigmatic cavity 3-4 mm wide at base, brownish (to olive-green) with a dark brown line across. FIG. 71.

13.6. *Ophrys argolica* subsp. *lesbis*

Subsp. *lesbis* is the most variable subspecies of *O. argolica* and may be confused with subsp. *lucis* – but see the key. Its most characteristic habitats include grassy garrigue and open groves and scrubs dominated by *Quercus pubescens* (downy oak), from sea level to 320 m altitude. Flowering takes place from mid-March to late April.

At present, subsp. *lesbis* is known from northwestern Lesbos and from a small area in southwestern Anatolia. Reports from Samos and Tilos should be checked, and a wider distribution on islands along the western coast of Anatolia seems likely. [Aeg; Ana].

DESCRIPTION. *Ophrys argolica* H. Fleischm. subsp. *lesbis* (Gölz & H. R. Reinhard) H. A. Pedersen & N. Faurholdt
Syn. *O. lesbis* Gölz & H. R. Reinhard.

Plant (8-)10-20(-28) cm tall with (1-)2-4(-6) flowers in a dense spike. **Sepals** whitish to rose-coloured or reddish violet, 13-18 × 7-10 mm, spreading. **Petals** velvety, 7.5-11 × 3.5-5.5 mm. **Lip** with (light yellowish) brown to olive-green ground colour on its central part, entire, widest above the middle, 13-16 × 13.5-20 mm, the basal half shaggy of fairly short, whitish to purplish brown hairs along the margin; terminal point triangular (1-2 mm wide at base); **mirror** bluish grey to reddish violet with or without a pale border, without connection to the base of the lip. Stigmatic cavity 2-4 mm wide at base, brownish with pale centre. FIG. 72.

REFERENCES: Baumann & Künkele (1982, 1984, 1988), Buttler (1986), E. Danesch & O. Danesch (1969), Davies et al. (1983), Del Prete & Tosi (1988), Delforge (1994, 2001, 2005), Dimitrov et al. (2001), Hansson (2001), Kalteisen & Reinhard (1987), Keitel & Remm (1991), Nelson (1962), Paulus & Gack (1990b, 1999), Rossi (2002), Selisky (2002), Sundermann (1980).

14. *Ophrys ferrum-equinum*

The Latin name "ferrum-equinum" refers to the mirror, which is usually shaped like a horseshoe. *Ophrys ferrum-equinum* is a beautiful and richly coloured bee orchid that one can often enjoy, as it is common in major parts of mainland Greece and on several Ionian and Aegean islands. This species has a superficial resemblance to *O. sphegodes* subsp. *spruneri* which, however, can be readily recognised by its usually H-shaped **mirror** with distinct connections to the base of the lip. In the islands of Ikaria and Naxos, deviant plants of *O. ferrum-equinum* are seen which some authors recognise as a separate species, *O. icariensis* M. Hirth & H. Spaeth. In Naxos, we have observed such plants growing side by side with typical subsp. *ferrum-equinum*, and we think that they are hybrids with the latter constituting one of the parental taxa.

The morphological variation in this species mainly falls between two

MAP 16. The European range of *Ophrys ferrum-equinum*. Two subspecies are recognized in Europe; only in southwestern Greece do their geographic ranges overlap.

groups of populations which we have decided to treat as separate subspecies. Subsp. *ferrum-equinum* is pollinated by the megachilid bee *Chalicodoma parietina*, but the identity of the pollinator of subsp. *gottfriediana* still remains to be revealed.

HABITAT. Calcareous, dry to moist soil in full sunlight to light shade, from sea level to 1000 m altitude. Typical habitats include roadside slopes, stony grassland, garrigue, open pine woods and pesticide-free olive groves.
FLOWERING TIME. From March to May, with a peak from mid-March to mid-April.
DISTRIBUTION. From the Balkans to Anatolia, where by far the major number of populations are situated in the southwest. There are no confirmed finds from Crete, but the species occurs in Karpathos. Additionally, it seems likely that *O. ferrum-equinum* occurs in Bulgaria – see note under subsp. *ferrum-equinum*. [Aeg, Alb, Cre, Gre; Ana] MAP 16.
DESCRIPTION. *Ophrys ferrum-equinum* Desf.
Synonyms are listed under the subspecies.

Plant compact to slender, 10–35 cm tall with 2–8 flowers in a lax spike.
Sepals violet to (greenish) white or olive-green, ovate to oblanceolate-
oblong, 10–17 × 3.5–9 mm; dorsal sepal nearly flat, more or less incurved,
from the base reflexed. **Petals** rose-coloured to purplish violet or white with
a greenish tinge, narrowly triangular to lanceolate-oblong with flat or wavy
margins, 6–11 × 2.5–4 mm, (nearly) glabrous, recurved to spreading. **Lip**
with (purplish) black ground colour, straight, flat or with more or less
recurved sides, entire to moderately three-lobed, 10–18.5 × 10–19.5 mm,
dark velvety along the margin (otherwise nearly glabrous); bulges rarely
present and then only weakly developed, distinctly isolated from the margin
of the lip (side lobes flat, if present); front edge rounded to slightly
emarginate, often provided with a short, porrect to downward directed,
triangular to awl-shaped point; **mirror** distinct, consisting of a more or less
horseshoe-shaped figure or two longitudinal bands of drop-shaped spots,
placed centrally without connection to the base of the lip, shining bluish
grey (rarely with a white border). **Column** acute (to obtuse), not tapering
towards the base (in side view); stigmatic cavity at least as wide as long and
approximately twice as wide as the anther, with dark, lateral, eye-like knobs
at base. FIGS 73–74.

Key to the subspecies

1. Sepals bright rose-coloured to violet (rarely white or
   greenish). Distal part of the lip with spreading to slightly
   recurved sides, making the lip appear elliptic to obovate and
   widest in its distal part when viewed from above . . . .
   . . . . 14.1. **subsp. *ferrum-equinum***
1. Sepals olive-green to white or muddy rose-coloured (rarely
   violet). The distal part of the lip with completely reflexed
   sides, making the lip appear triangular and widest in its basal
   part when viewed from above . . . . 14.2. **subsp. *gottfriediana***

14.1. *Ophrys ferrum-equinum* subsp. *ferrum-equinum*

Normally, the two subspecies of *O. ferrum-equinum* are well distinguished
and easily recognised, but, probably due to hybridisation and backcrossing,
confusing intermediary forms frequently occur in places where both
subspecies are present. Additionally, aberrant individuals with strongly
recurved lip sides (almost as in subsp. *gottfriediana*) are occasionally
encountered in populations of subsp. *ferrum-equinum* (e.g. in Lesbos, Samos
and Rhodes). Anatolian populations of large-flowered plants with
downward directed lips were recently described as a separate species, *O.*

FIG. 73. *Ophrys ferrum-equinum* subsp. *ferrum-equinum* (**A.** Greece, Rhodes, Isidoros, 17th April 1996; **B-C.** Greece, Chios, Eleovounos, 16th April 2005). Photos by H. Æ. Pedersen.

labiosa Kreutz. We have observed similar individuals in populations of *O. ferrum-equinum* subsp. *ferrum-equinum* from Lesbos, Chios and Samos. The flowering season is the same as indicated under the species.

The distribution covers the whole range of the species. A recent record of *O. argolica* from Bulgaria seems referable to *O. ferrum-equinum* subsp. *ferrum-equinum*, but this supposition needs to be verified. [Aeg, Alb, Cre, Gre; Ana].

DESCRIPTION. *Ophrys ferrum-equinum* Desf. subsp. *ferrum-equinum*.

Sepals bright rose-coloured to violet (rarely white or greenish), 10–17 × 5–9 mm. **Petals** rose-coloured to purplish violet, 6–11 × 2.5–4 mm. **Lip** 10–17 × 10–19 mm; its distal part with spreading to slightly recurved sides, making the lip appear elliptic to obovate and widest in its distal part when viewed from above. FIG. 73.

169

14.2. *Ophrys ferrum-equinum* subsp. *gottfriediana*

Subsp. *gottfriediana* is little variable and easy to identify under normal conditions. However, hybrids with subsp. *ferrum-equinum* and aberrant forms of the latter may cause difficulties (cf. the discussion under subsp. *ferrum-equinum*). The flowering season is the same as indicated under the species.

Subsp. *gottfriediana* is endemic to Greece, where it is known with certainty from Ithaka, Kefallinia, Zakinthos and other islands in the Ionian Sea as well as from adjoining parts of the mainland. Recent reports from islands in the Aegean Sea should be checked. [Gre].

DESCRIPTION. *Ophrys ferrum-equinum* Desf. subsp. *gottfriediana* (Renz) E. Nelson
Syn. *O. gottfriediana* Renz.

Sepals olive-green to white or muddy rose-coloured (rarely violet), 11-15 × 3.5-6 mm. **Petals** greenish white to rose-coloured or purplish violet, 7-9 × 2.5-3 mm. **Lip** 11.5-18.5 × 10.5-19.5 mm; its distal part with completely reflexed sides, making the lip appear triangular and widest in its basal part when viewed from above. FIG. 74.

FIG. 74

FIG. 74. *Ophrys ferrum-equinum* subsp. *gottfriediana* (Greece, the Peloponnese, Adriani, 17th April 1997). Photo by A. Gøthgen.

REFERENCES: Baumann & Künkele (1982, 1988), Buttler (1986), E. Danesch & O. Danesch (1969), Davies et al. (1983), Delforge (1994, 2001, 2005), Dimitrov et al. (2001), Kretzschmar & Kretzschmar (2003), Nelson (1962), Paulus & Gack (1990b, 1992), Sundermann (1980), Vöth (1984).

15. *Ophrys bertolonii*

Ophrys bertolonii is remarkable for its very dark and markedly saddle-shaped lip which is often bent at nearly right angles. The species is pollinated by the megachilid bees *Chalicodoma parietina* and *C. pyrenaica*. It shares the former with, for example, the western Mediterranean *Ophrys atlantica*, and the saddle-shaped lip in both bee orchids probably represents a common, but independent, adaptation to this particular pollinator. *Ophrys bertolonii* exhibits little variation and can hardly be confused with anything but representatives of the partly stabilised hybrid complex between this species and *O. sphegodes* (see H5. *O. ×flavicans*). *Ophrys ×flavicans*, however, has less (if at all) saddle-shaped lip and a column that is not tapering towards the base (in side view). Furthermore, its stigmatic cavity is only about as long as wide.

HABITAT. Usually in dry to moist, calcareous soil in full sunlight, from sea level to 1200 m altitude. Typical habitats include roadside verges, garrigue, olive groves and grassland as well as open places in forest and maquis.

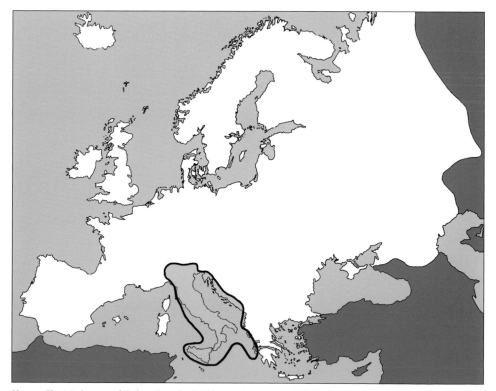

MAP 17. The total range of *Ophrys bertolonii*. This species is not subdivided into subspecies.

FIG. 75A

FIG. 75. *Ophrys bertolonii* (**A-B**. Italy, Sicily, Buccheri, 17th April 1999; **C**. Italy, Puglia, Alberobello, 20th April 2002; **D**. Italy, Toscana, Alberese, 10th April 2001). Photos by H. Æ. Pedersen (**A-B**), N. Faurholdt (**C-D**).

FLOWERING TIME. From (March-)April to June, with a peak from mid-April to mid-May.

DISTRIBUTION. Central Mediterranean; mainly occuring in Italy, from the flood plain of the river Po south to Sicily. In the Balkans, it ranges from Istria in Croatia to Montenegro and also occurs on the Ionian island of Corfu. Reports from France seem referable to *O.* ×*flavicans*. [Gre, Ita, Sic, Yug] MAP 17.

DESCRIPTION. *Ophrys bertolonii* Moretti.

Plant slender to relatively compact, 10-35 cm tall with 2-8 flowers in a lax spike. **Sepals** rose-coloured to (greenish) white, narrowly elliptic to (ob)lanceolate-oblong, 13-18.5 × 6-8 mm; dorsal sepal shallowly boat-shaped to nearly flat, more or less incurved, from the base reflexed. **Petals** violet to rose-coloured, narrowly oblong with flat margins, 8-12 × 2-4.5 mm,

FIG. 75B

FIG. 75C

FIG. 75D

pubescent, spreading to slightly incurved. **Lip** with blackish brown ground colour, markedly saddle-shaped with recurved sides, entire, 12.5-19 × 14-17 mm, shaggy along the margin, velvety on the central part (except on the mirror); bulges absent; front edge emarginate, provided with a short, erect, awl-shaped to narrowly triangular point; **mirror** distinct, transversely elliptic to nearly square, placed on the distal part of the lip without connection to its base, shining greyish blue to greyish violet. **Column** acute, distinctly tapering towards the base (in side view); stigmatic cavity longer than wide and slightly wider than the anther, with dark, lateral, eye-like knobs at base. FIG. 75.

REFERENCES: Baumann & Künkele (1982, 1988), Büel (1978), Buttler (1986), E. Danesch & O. Danesch (1969), Davies et al. (1983), Del Prete & Tosi (1988), Delforge (1994, 2001, 2005), Gulli et al. (2003), Nelson (1962), Paulus (2000), Paulus & Gack (1986), Rossi (2002), Souche (2004), Sundermann (1980).

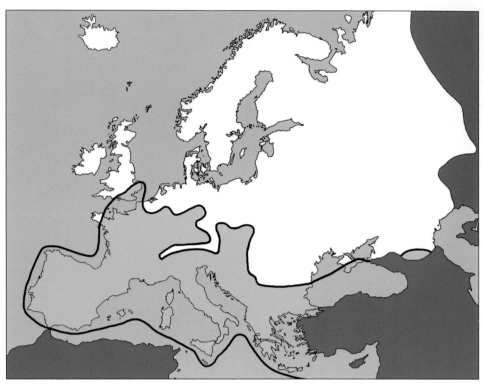

Map 18. The European range of *Ophrys sphegodes*. Twelve subspecies are recognized in Europe, where the highest diversity is found in Greece (eight subspecies).

16. *Ophrys sphegodes*

Ophrys sphegodes is one of the most widespread and variable bee orchids, displaying a diverse and confusing array of forms in major parts of central and southern Europe; we recognise more subspecies of *O. sphegodes* than of any other *Ophrys* species. On top of this, *O. sphegodes* is involved in partly stabilised hybrid complexes with *O. fuciflora* (see H4. *O. ×arachnitiformis*) and *O. bertolonii* (see H5. *O. ×flavicans*). We cannot briefly explain how to distinguish *O. sphegodes* from the other European bee orchids, but recommend frequent use of the Key to the species and partly stabilised hybrid complexes p. 62.

The complex morphological variation mainly falls between 12 groups of populations which also differ from each other with regard to distribution, flowering season and/or habitat preferences. Against this background we recognise them as 12 subspecies. It is evident that pollinator specificity also contributes to keeping the subspecies separate. *Ophrys sphegodes* is pollinated almost entirely by a number of (for each subspecies often specific) andrenid bees of the genus *Andrena*. The only consistent exceptions are the colourful subsp. *sipontensis* and subsp. *spruneri* from Italy and Greece, respectively. Both of these subspecies are pollinated by the xylocopid bee *Xylocopa iris*.

HABITAT. Dry to moist soil in full sunlight to light shade, from sea level to 1400 m altitude. Typical habitats include garrigue, roadside slopes, open woods, grassland and pesticide-free olive groves on calcareous as well as slightly acid ground.

FLOWERING TIME. From February to June(-July), in most places with a peak in March–April.

DISTRIBUTION. From southern England across central and southern Europe and the Levant to the Caucasus and northern Iran. [Aeg, Alb, Aus, Bal, Bel, Bul, Cor, Cre, Cze, Eng, Fra, Ger, Gre, Hun, Lux, Ita, Mal, Por, Rum, Rus, Sar, Sic, Spa, Swi, Tur, Ukr, Yug; Ana, Cyp, Isr] MAP 18.

DESCRIPTION. *Ophrys sphegodes* Mill.

Synonyms are listed under the subspecies.

Plant slender, (10-)15–60(-70) cm tall with 2-12(-18) flowers in a lax to dense spike. **Sepals** pale green to yellowish green, (olive-)green, white or purplish violet (the lateral ones sometimes distinctly bicoloured with the mid-vein constituting a boundary), now and then suffused with violet to purplish brown, ovate to elliptic or (ob)lanceolate-oblong, 7-19 × (2-)3-9 mm; dorsal sepal nearly flat (less often shallowly boat-shaped), more or less incurved, from the base reflexed. **Petals** (ochre-)yellow to olive-green or pale green (often suffused with red to brown) or bright purplish violet to ruby, linear-lanceolate to oblong or (ovate-)triangular with flat to strongly wavy margins, 4-13 × 1.5-7 mm, glabrous to pubescent, recurved to spreading. **Lip** with (blackish, reddish or purplish) brown ground colour and often a light brown to yellow or yellowish green, less frequently reddish brown margin, straight, nearly flat or with more or less recurved sides, entire or slightly to moderately (rarely deeply) three-lobed close to the middle, 5-18(-20) × 7-18 mm, (sub)glabrous to strongly shaggy along the margin (often more hairy towards the base), otherwise (sub)glabrous; bulges weakly developed to obliquely conical (sometimes absent), distinctly isolated from the margin of the lip (side lobes flat, if present); front edge obtuse to emarginate, usually provided with a short, porrect to downward directed, triangular point; **mirror** distinct absent in subsp. *helenae*, most frequently consisting of an H-shaped figure or another simple figure derived from this, sometimes considerably more complicated and/or marbled, distinctly connected to the base of the lip, dull greyish blue to violet (rarely reddish brown), sometimes with a pale border. **Column** acute (to obtuse); not tapering towards the base (in side view); stigmatic cavity approximately as wide as long and approximately twice as wide as the anther, with dark or pale, lateral, eye-like knobs at base. FIGS 76-87.

Key to the subspecies

1. Bulges of the lip well-developed, obliquely conical 2
1. Bulges of the lip absent or very low and broadly rounded 6

2. Lip with a thick, marginal, shaggy border of strikingly long hairs all around . . . . 16.3. **subsp.** *atrata*
2. At least the distal half of the lip without strikingly long and dense hairiness along the margin 3

3. Lip wedge-shaped at base . . . . 16.12. **subsp.** *gortynia*
3. Lip rounded to truncate at base 4

4. Lip 5-9 mm long . . . . 16.11. **subsp.** *cretensis*
4. Lip 9-18(-20) mm long. 5

5. Lateral sepals of one colour (green to white). Each of the two bulges of the lip at most one fourth as large as the stigmatic cavity; front edge of the lip usually emarginate around a small terminal point . . . . 16.1. **subsp.** *sphegodes*
5. Lateral sepals usually bicoloured (green and purplish brown) with the mid-vein constituting a boundary. Each of the two bulges of the lip at least half as large as the stigmatic cavity; front edge of the lip rounded or acuminate with a small terminal point . . . . 16.10. **subsp.** *mammosa*

6. Sepals purplish violet to white (very rarely light greenish). Petals bright purplish violet to ruby 7
6. Sepals green (to white), occasionally with purplish brown markings. Petals (ochre-)yellow to olive-green or pale green (often suffused with red or brown) 8

7. Lateral sepals usually distinctly bicoloured with the mid-vein constituting a boundary. Lip three-lobed (often deeply so, very rarely subentire) . . . . 16.6. **subsp.** *spruneri*
7. Lateral sepals never distinctly bicoloured with the mid-vein constituting a boundary. Lip entire . . . . 16.5. **subsp.** *sipontensis*

8. Mirror absent or, less frequently, very obscure (normally, it can only be recognised in backlight) . . . . 16.7. **subsp.** *helenae*
8. Mirror distinct, often H-shaped 9

9. Basal part of the lip shortly velvety to subglabrous along the margin; front edge of the lip rounded to acuminate with a short terminal point 10
9. Basal part of the lip more or less shaggy along the margin;

front edge of the lip usually distinctly emarginate around a
short terminal point 12

10. Lip 5-9 mm long; the distal half without (or, less frequently,
with a no more than 1 mm wide) light brown to yellow
margin . . . . 16.11. **subsp.** *cretensis*
10. Lip 9-14 mm long; the distal half with a 1.5-3 mm wide,
yellow to yellowish green, rarely reddish brown margin 11

11. Stigmatic cavity speckled green/brown. The eye-like
knobs of the column pale (yellowish) green . . . .
. . . . 16.9. **subsp.** *aesculapii*
11. Stigmatic cavity uniformly brown (or very nearly so).
The eye-like knobs of the column greyish blue . . . .
. . . . 16.8. **subsp.** *epirotica*

12. Lip with dark brown ground colour; mirror basically
H-shaped, but usually with two additional short arms from
the base . . . . 16.4. **subsp.** *passionis*
12. Lip with medium brown ground colour; mirror H-shaped
(occasionally more complicated or marbled), only in very
rare cases with two additional well-defined arms from the base 13

13. Plant slender. Lip 10-16 mm long, longer than the dorsal
sepal, without or with only a narrow (c. 1 mm wide) yellow
margin . . . . 16.1. **subsp.** *sphegodes*
13. Plant robust. Lip 6.5-10 mm long, shorter than the dorsal
sepal, with or without an up to 2 mm wide yellow or
yellowish green margin . . . . 16.2. **subsp.** *litigiosa*

16.1. *Ophrys sphegodes* subsp. *sphegodes*

Subsp. *sphegodes* is highly variable and may be confused with, in particular
subspecies *passionis* and *litigiosa* – but see the key. The variation within
subsp. *sphegodes* is particularly pronounced in Italy, southern France and the
northern part of the Balkans. Flowering takes place from March to June. In
the Apennines, however, flowering individuals are still encountered in July.

The distribution encompasses the whole western part of the specific
range, east to Hungary, former Yugoslavia, Albania and Corfu. [Alb, Aus,
Bel, Cze, Eng, Fra, Ger, Gre, Hun, Ita, Lux, Sar, Sic, Spa, Swi, Yug].

DESCRIPTION. *Ophrys sphegodes* Mill. subsp. *sphegodes*
Syn. *O. aranifera* Hudson; *O. cephalonica* (B. & H. Baumann) J. Devillers-
Terschuren & P. Devillers; *O. sphegodes* Mill. subsp. *cephalonica* B. & H.
Baumann; *O. cilentana* J. Devillers-Terschuren & P. Devillers; *O. classica* J.
Devillers-Terschuren & P. Devillers; *O. exaltata* Ten. s.s.; *O. fuciflora* (F. W.

FIG. 76A

FIG. 76B

Schmidt) Moench subsp. *exaltata* (Ten.) E. Nelson, nom. tant.; *O. holoserica* (Burm.f.) Greuter subsp. *exaltata* (Ten.) Landwehr, comb. inval., nom. tant.; ?*O. hebes* (Kalopissis) B. & E. Willing; ?*O. sphegodes* Mill. subsp. *hebes* Kalopissis; *O. liburnica* P. Devillers & J. Devillers-Terschuren; ?*O. majellensis* (Daiss.) P. Delforge; *O. montenegrina* (H. Baumann & Künkele) J. Devillers-Terschuren & P. Devillers; *O. sphegodes* Mill. subsp. *montenegrina* H. Baumann & Künkele; ?*O. negadensis* G. & W. Thiele; *O. panormitana* (Tod.) Soó [non (Tod.) Landwehr, comb. superfl.] s.s.; *O. sphegodes* Mill. subsp. *panormitana* (Tod.) E. Nelson; *O. panormitana* (Tod.) Soó subsp. *praecox* (Corrias) Paulus & Gack; *O. panormitana* (Tod.) Soó var. *praecox* (Corrias) P. Delforge; *O. sphegodes* Mill. subsp. *praecox* Corrias; ?*O. provincialis* (H. Baumann & Künkele) Paulus; ?*O. sphegodes* Mill. subsp. *provincialis* H. Baumann & Künkele; *O. sphegodes* Mill. subsp. *provincialis* E. Nelson, nom. inval.; *O. sphegodes* Mill. subsp. *sicula* E. Nelson, nom. inval.; *O. tarquinia* P. Delforge.

Plant slender, (10–)15–40(–60) cm tall with (2–)3–9(–15) flowers in a lax spike. **Sepals** green to white (occasionally with a yellowish or rose-coloured

FIG. 76C

Fig. 76. *Ophrys sphegodes* subsp. *sphegodes* (**A.** Italy, Sicily, Bosco di Ficuzza, 21st April 1999; **B-C.** France, Meuse, Montmédy 9th May 1992). Photos by H. Æ. Pedersen (**A**), C. A. J. Kreutz (**B-C**).

tinge), 8-14 × (2-)3-7 mm. **Petals** yellowish to olive-green (sometimes suffused with rust-red), 4-10(-11) × 2-4.5 mm, approximately half as wide as the sepals. **Lip** with medium brown ground colour, occasionally provided with a c. 1 mm wide yellow margin, entire (rarely more or less three-lobed) with rounded to truncate base, 9.5-16 × 9-18 mm, longer than the dorsal sepal, margin of the basal part more or less shaggy, margin of the distal part velvety to subglabrous; front edge emarginate (rarely rounded) around a short terminal point; bulges almost absent to obliquely conical (but each of them rarely more than one fourth as large as the stigmatic cavity); **mirror** H-shaped (rarely a little more complicated). Eye-like knobs of the column dark. FIG. 76.

16.2. *Ophrys sphegodes* subsp. *litigiosa*

This subspecies is fairly variable and may easily be confused with subsp. *sphegodes* – but see the key. In certain parts of Italy, the floral variation is so pronounced that some authors recognise Tyrrhenian populations as a separate species, *O. argentaria*. This entity, however, is not very well-defined; neither its flowering season nor habitats differ from those of typical

O. sphegodes subsp. *litigiosa*, and pollinator specificity has not been unequivocally demonstrated. For these reasons, we merely assign varietal rank to the Tyrrhenian entity. The flowering season is relatively early, in the Mediterranean peaking from mid–March to mid–April. Further north, flowering plants can be found in May and early June.

This subspecies is mainly distributed in parts of France, in central Italy, and in the northwestern part of the Balkan Peninsula. To the west and the north, it reaches Spain (Cataluña and possibly Rioja) and central Germany, respectively. [Fra, Ger, Ita, Spa, Swi, Yug].

DESCRIPTION. *Ophrys sphegodes* Mill. subsp. *litigiosa* (E. G. Camus) Bécherer Syn. [var. *litigiosa*]: *O. araneola* Rchb.; *O. argensonensis* Guérin & Merlet; *O. ausonia* P. Devillers et al.; ?*O. illyrica* S. & K. Hertel; *O. incantata* P. Devillers & J. Devillers-Terschuren; *O. litigiosa* E. G. Camus; ?*O. quadriloba* (Rchb.f.) E. G. Camus; *O. tommasinii* Vis.; *O. virescens* Philippe ex Gren. Syn. [var. *argentaria* (J. Devillers-Terschuren & P. Devillers) N. Faurholdt]: *O. argentaria* J. Devillers-Terschuren & P. Devillers.

FIG. 77A

FIG. 77B

FIG. 77. *Ophrys sphegodes* subsp. *litigiosa*. **A-B.** var. *litigiosa*; **C-D.** var. *argentaria* (**A-B.** Croatia, Istria, Bale, 16th May 2004; **C-D.** Italy, Toscana, Monte Argentario, 8th April 2001). Photos by N. Faurholdt.

FIG. 77C

FIG. 77D

Plant robust, (10-)15-45 cm tall with 2-10 (-15) flowers in a lax to dense spike. **Sepals** (yellowish) green to white (occasionally with a rose-coloured tinge), 7-12 × 4-7 mm. **Petals** olive-green to ochre-yellow (occasionally suffused with rust-red), 5-8 × 2-4.5 mm, at least half as wide as the sepals. **Lip** with medium brown ground colour, often provided with an up to 2 mm wide, yellow to yellowish green margin, entire (rarely slightly three-lobed) with rounded to truncate base, 6.5-9.5 × 7.5-11.5 mm, shorter than the dorsal sepal, margin of the basal part more or less shaggy, margin of the distal part velvety to subglabrous; front edge emarginate around a short terminal point; bulges absent or only weakly developed; **mirror** H-shaped to more complicated or marbled. Eye-like knobs of the column dark. FIG. 77.

Var. *litigiosa* is mainly characterised by its small, H-shaped mirror, which is rarely delimited by a white border, and by the up to 2 mm wide yellow (to yellowish brown) margin of its lip. – **Plant** (10-)20-45 cm tall with (2-)4-10(-15) flowers in a dense to relatively lax spike. **Sepals** (yellowish) green to white. **Petals** of the same colour as the sepals or a little darker, of one colour or with slightly darker margins. **Lip** with an up to c. 2 mm wide yellow (to yellowish brown) margin; **mirror** small, H-shaped, usually without a white border. [FIG. 77A-B]. Var. *litigiosa* occurs throughout the geographic range of the subspecies. [Fra, Ger, Ita, Spa, Swi, Yug].

Var. *argentaria* is mainly characterised by its fairly large and complicated (often somewhat marbled) mirror, which is usually delimited by a white border, and by the quite narrow, usually light brownish to yellowish, short-hairy margin of its lip. – **Plant** 15-35 cm tall with 2-10 flowers in a lax spike. **Sepals** light green. **Petals** olive-green to ochre-yellow, sometimes with rust-red margins. **Lip** with a quite narrow, light brown to yellow or yellowish green margin; **mirror** fairly large and complicated (often including, e.g., 1-3 eye-like spots), often somewhat marbled and usually delimited by a white border. [FIG. 77C-D]. Up to now, this variety has only been found with certainty along the west coast of mainland Italy, from Pisa to Rome (primarily in the provinces of Grosetto and Livorno). Similar plants from northern Spain should possibly also be referred to this variety. [Ita].

16.3. *Ophrys sphegodes* subsp. *atrata*

Subsp. *atrata* exhibits very little variation and is easy to identify. A number of populations in Toscana, however, consist of individuals devoid of the, otherwise characteristic shaggy border of the lip. Flowering takes place from March to May, but generally a little later than in subsp. *sphegodes*.

Subsp. *atrata* is distributed in central and western South Europe, from Portugal to the western part of the Balkan Peninsula. It is particularly common in the greater islands and in the southern part of mainland Italy. [Alb, Bal, Cor, Fra, Ita, Mal, Por, Sar, Sic, Spa, Yug].

DESCRIPTION. *Ophrys sphegodes* Mill. subsp. *atrata* (Lindl.) E. Mayer
Syn. *O. atrata* Lindl.; ?*O. brutia* P. Delforge; *O. incubacea* Bianca.

Plant relatively slender, 20-40(-60) cm tall with 2-8 flowers in a lax spike. **Sepals** green (occasionally with a rose-coloured tinge), 10-15.5 × 4-7.5 mm. **Petals** (olive-)green to muddy ochre-yellow, 6.5-9 × 2.5-5 mm, approximately half as wide as the sepals. **Lip** with dark brown to blackish brown ground colour, entire (rarely slightly three-lobed) with rounded to truncate base, (8-)10-14 × (8-)10-14.5 mm, shorter than the dorsal sepal, with a thick, marginal, shaggy border of strikingly long hairs all around; front edge emarginate around a short terminal point; bulges obliquely conical (but each of them at most half as large as the stigmatic cavity); **mirror** H-shaped. Eye-like knobs of the column dark. FIG. 78.

16.4. *Ophrys sphegodes* subsp. *passionis*

Variation within subsp. *passionis* is modest, and this subspecies can hardly be confused with anything but aberrant individuals of subsp. *sphegodes*. In mixed colonies, hybridisation may cause additional problems. Throughout most of the geographic range, flowering takes place from March to May, with a peak in April. In higher mountains, however, flowering plants may still be found in June(-July).

The distribution is insufficiently known, but it seems to encompass northeastern Spain, southern and westernmost France, Sardinia (where this subspecies is very rare), Sicily and mainland Italy from Toscana to Calabria. [Fra, Ita, Sar, Sic, Spa].

FIG. 78. *Ophrys sphegodes* subsp. *atrata* (Italy, Sicily, Buccheri, 17th April 1999). Photo by H. Æ. Pedersen.

183

FIG. 79. *Ophrys sphegodes* subsp. *passionis* (**A.** Italy, Sicily, Castiglione, 16th April 1992; **B.** Italy, Puglia, Monte Gargano, 16th April 1995). Photos by N. Faurholdt (**A**), H. Æ. Pedersen (**B**).

FIG. 79B

FIG. 79A

DESCRIPTION. *Ophrys sphegodes* Mill. subsp. *passionis* (Sennen ex J. Devillers-Terschuren & P. Devillers) Sanz & Nuet
Syn. *O. garganica* O. & E. Danesch [non (E. Nelson) O. & E. Danesch, comb. inval.]; *O. sphegodes* Mill. subsp. *garganica* E. Nelson, nom. inval.; *O. passionis* Sennen ex J. Devillers-Terschuren & P. Devillers var. *garganica* (O. & E. Danesch) P. Delforge; *O. passionis* Sennen ex J. Devillers-Terschuren & P. Devillers s.s.

Plant relatively slender, 20-40(-45) cm tall with 4-8 flowers in a lax spike. **Sepals** green, 10-14 × 4-7 mm. **Petals** olive-green to ochre-yellow (occasionally suffused with rust-red), 8-11 × 3.5-7 mm, nearly as wide as the sepals. **Lip** with dark brown ground colour, often provided with a light reddish brown (less frequently yellow) margin, entire (rarely slightly three-lobed) with rounded to truncate base, 8-14 × 13-17 mm, approximately as long as the dorsal sepal, the basal part more or less shaggy along the margin, the distal part velvety to subglabrous along the margin; front edge emarginate around a short terminal point; bulges absent or weakly developed; **mirror** basically H-shaped, but usually with two additional short arms from the base. Eye-like knobs of the column dark. FIG. 79.

16.5. *Ophrys sphegodes* subsp. *sipontensis*

This subspecies is easily recognised and little variable. It grows on poor, dry to humid grassland and in garrigue (prevailingly in places dominated by asphodels), up to 600 m altitude. Flowering takes place from March to May, with a peak in April.

Subsp. *sipontensis* has an extremely narrow distribution, being restricted to the Siponte plateau on the southern slope of Monte Gargano in Puglia. [Ita].

DESCRIPTION. *Ophrys sphegodes* Mill. subsp. *sipontensis* (O. & E. Danesch) H. A. Pedersen & N. Faurholdt [non Gumpr., nom. inval.]
Syn. *O. sipontensis* O. & E. Danesch [non (Gumpr.) O. & E. Danesch, comb. inval. et non R. Lorenz & C. Gembardt, nom. illeg.]; *O. garganica* O. & E. Danesch subsp. *sipontensis* (Gumpr.) Del Prete, comb. inval.

Plant relatively slender, (15-)20-50(-60) cm tall with 2-8 flowers in a lax spike. **Sepals** purplish violet to white, 10-15 × 4-7 mm. **Petals** bright

FIG. 80

FIG. 80. *Ophrys sphegodes* subsp. *sipontensis* (Italy, Puglia, Monte Gargano, 11th April 1995). Photo by H. Æ. Pedersen.

185

purplish violet to ruby, 7-12 × 3-7 mm, usually almost as wide as the sepals. **Lip** with dark brown to blackish brown ground colour, entire with rounded to truncate base, 10-15 × 10-17 mm, approximately as long as the dorsal sepal, the basal part (and less pronouncedly the distal part) shaggy along the margin; front edge emarginate around a short terminal point; bulges absent or weakly developed; **mirror** H-shaped (rarely a little more complicated) or consisting of two parallel longitudinal bands. Eye-like knobs of the column dark. FIG. 80.

16.6. *Ophrys sphegodes* subsp. *spruneri*

Subsp. *spruneri* is fairly variable in the shape of the lip and in the colour of sepals and petals. Where it grows together with *O. sphegodes* subsp. *mammosa* or *O. ferrum-equinum*, identification can also be hampered by hybridisation. Flowering takes place from late February to early May, with a peak in March and April.

Subsp. *spruneri* is endemic to Greece, where it occurs in the mainland as well as in Crete and on some of the smaller islands of the Aegean (east to Ikaria and Chios). [Aeg, Cre, Gre].

DESCRIPTION. *Ophrys sphegodes* Mill. subsp. *spruneri* (Nyman) E. Nelson Syn. *O. grigoriana* G. & H. Kretzschmar; *O. spruneri* Nyman subsp. *grigoriana* (G. & H. Kretzschmar) H. Kretzschmar; *O. sphaciotica* H. Fleischm.; *O. spruneri* Nyman s.s.

Plant slender to relatively robust, (10-)15-40(-50) cm tall with 2-8 flowers in a lax (to dense) spike. **Sepals** purplish violet to white (very rarely light greenish), the lateral ones usually distinctly bicoloured (white/purplish violet) with the mid-vein constituting a boundary, 10-16.5 × (3-)5-6.5 mm. **Petals** bright purplish violet to ruby, (5-)8-11 × 2-4 mm, approximately half as wide as the sepals. **Lip** with blackish brown ground colour, often provided with a paler margin, three-lobed (often deeply so, very rarely subentire) with rounded to truncate base, 10-15 × (10-)12-18 mm, approximately as long as the dorsal sepal, the basal part velvety along the margin, the distal part (sub)glabrous along the margin; front edge rounded with a short terminal point; bulges absent or weakly developed; **mirror** H-shaped (rarely a little more complicated) or consisting of two parallel longitudinal bands. Eye-like knobs of the column dark. FIG. 81.

FIG. 81. *Ophrys sphegodes* subsp. *spruneri* (Greece, Crete, Rodovani, 23rd March 1997). Photo by N. Faurholdt.

16.7. *Ophrys sphegodes* subsp. *helenae*

This subspecies is very characteristic and can hardly be confused with anything but monstrous forms of other bee orchids (forms devoid of a mirror). Flowering takes place from April to June, with a peak in April and May.

The distribution of subsp. *helenae* encompasses southern Albania and northern Greece. Additionally, one reliable report from the Peloponnese exists. [Alb, Gre].

DESCRIPTION. *Ophrys sphegodes* Mill. subsp. *helenae* (Renz) Soó & D. M. Moore
Syn. *O. helenae* Renz.

Plant robust, 15-40 cm tall with 2-8 flowers in a dense to relatively lax spike. **Sepals** pale green to yellowish green, often more or less suffused with violet, 11-15.5 × 5-8 mm. **Petals** pale green to ochre-yellow (occasionally suffused with violet or rust-red), 6-13 × 2-4 mm, approximately half as

FIG. 82

FIG. 82. *Ophrys sphegodes* subsp. *helenae* (Greece, Epiros, Igoumenitsa, 11th April 1997). Photo by A. Gøthgen.

FIG. 83. *Ophrys sphegodes* subsp. *epirotica* (**A-B**: Greece, Epiros, Metsovon, 22nd May 1997). Photos by A. Gøthgen.

wide as the sepals. **Lip** with purplish brown ground colour, entire with rounded to truncate base, 11–18 × 15–23 mm, usually longer than the dorsal sepal, the basal part shortly velvety to subglabrous along the margin, the distal part (sub)glabrous along the margin; front edge emarginate with a short terminal point; bulges absent or weakly developed; **mirror** absent or very obscure (it can normally only be recognised in backlight). Eye-like knobs of the column strongly reduced, dark. FIG. 82.

16.8. *Ophrys sphegodes* subsp. *epirotica*

This subspecies is reminiscent of the earlier flowering subsp. *aesculapii* and may be confused with it, but see the key. Subsp. *epirotica* flowers relatively late, from late April to mid-June.

The distribution encompasses Albania and northwestern Greece. [Alb, Gre].

DESCRIPTION. *Ophrys sphegodes* Mill. subsp. *epirotica* (Renz) Gölz & H. R. Reinhard

Syn. *O. epirotica* (Renz) J. Devillers-Terschuren & P. Devillers.

Plant relatively slender, (15–)20–45 cm tall with 4–9(–15) flowers in a lax to relatively dense spike. **Sepals** yellowish green to olive-green, sometimes with purplish brown markings, 10-14 × 4-7 mm. **Petals** bright yellowish green (occasionally suffused with rust-red), 6-9 × 2.5-4 mm, approximately half as wide as the sepals. **Lip** with light brown to blackish brown ground colour, provided with a broad, reddish brown to yellow or yellowish green margin, entire (rarely slightly three-lobed) with rounded to truncate base, 10-14 × 10-17 mm, approximately as long as the dorsal sepal, the basal part shortly velvety to subglabrous along the margin, the distal part (sub)glabrous along the margin; front edge rounded to shortly acuminate with a short terminal point; bulges absent or weakly developed; **mirror** H-shaped or consisting of two parallel longitudinal bands. Stigmatic cavity (almost) uniformly brown. Eye-like knobs of the column greyish blue. FIG. 83.

16.9. *Ophrys sphegodes* subsp. *aesculapii*

This subspecies is fairly constant and mainly varies with regard to the colour of the lip margin; occasionally, small- and few-flowered plants with a light brown lip margin are seen. Subsp. *aesculapii* may be confused with the later flowering subsp. *epirotica* – but see the key. Flowering takes place from mid-March to early May, with a peak in early April.

FIG. 84. *Ophrys sphegodes* subsp. *aesculapii* (**A.** Greece, Attika, Imettos near Athens, 30th March 1996; **B.** Greece, Attika, Imettos near Athens, 30th March 1991). Photos by N. Faurholdt (**A**), A. Gøthgen (**B**).

Subsp. *aesculapii* is endemic to Greece, where it is known from the southern and eastern part of the Peloponnese, from Euboea and from adjoining parts of the mainland. [Aeg, Gre].

DESCRIPTION. *Ophrys sphegodes* Mill. subsp. *aesculapii* (Renz) Soó
Syn. *O. aesculapii* Renz.

Plant relatively robust, 15-40 cm tall with 3-12 flowers in a relatively dense (to lax) spike. **Sepals** pale green to yellowish green, often with purplish brown markings, 9-14 × 3.5-5.5 mm. **Petals** yellowish green to olive-green (occasionally suffused with rust-red), 5-8.5 × 2-3.5 mm, approximately half as wide as the sepals. **Lip** with (dark) brown ground colour, provided with a broad, yellow to reddish brown margin, entire (rarely slightly three-lobed) with rounded to truncate base, 9-12 × 10-14 mm, approximately as long as the dorsal sepal, the basal part shortly velvety to subglabrous along the margin, the distal part (sub)glabrous along the margin; front edge rounded to shortly acuminate with a short terminal point; bulges weakly developed; **mirror** H-shaped. Stigmatic cavity speckled green/brown. Eye-like knobs of the column pale (yellowish) green. FIG. 84.

16.10. *Ophrys sphegodes* subsp. *mammosa*

Subsp. *mammosa* is highly variable as far as flowering time and floral morphology are concerned, and certain authors have in recent years proposed a considerable splitting of this entity, even at species level. In our opinion, however, the validity of this splitting is not sufficiently documented, for which reason we have decided to treat the whole complex as one subspecies. Within Europe, typical individuals of subsp. *mammosa* cannot be confused with any other bee orchid. On the other hand, hybrid swarms with, for example, subsp. *spruneri* may locally complicate identification. Individuals with a narrow yellow lip margin are most frequently encountered in the eastern part of the range. Flowering takes place from March to June.

Subsp. *mammosa* is distributed from the Balkans across the Levant to the Caucasus and northern Iran. [Aeg, Alb, Bul, Cre, Gre, Rus, Tur, Ukr, Yug; Ana, Cyp, Isr].

DESCRIPTION. *Ophrys sphegodes* Mill. subsp. *mammosa* (Desf.) Soó
Syn. *O. caucasica* Woronow ex Grossh.; *O. grammica* (B. & E. Willing) J. Devillers-Terschuren & P. Devillers; *O. herae* M. Hirth & H. Spaeth; *O. hystera* Kreutz & R. Peter; *O. leucophthalma* J. Devillers-Terschuren & P.

FIG. 85. *Ophrys sphegodes* subsp. *mammosa* (**A**. Greece, Rhodes, Isidoros, 17th April 1996; **B**. Greece, Samos, Zervos, 12th April 1999). Photos by H. Æ. Pedersen.

Devillers; *O. macedonica* (H. Fleischm. ex Soó) J. Devillers-Terschuren & P. Devillers; *O. mammosa* Desf. var. *macedonica* (H. Fleischm.) Landwehr, comb. inval.; *O. mammosa* Desf. s.s.; *O.* ×*pseudomammosa* auct. [non Renz].

Plant robust to relatively slender, (15-)20-60(-70) cm tall with 2-12(-18) flowers in a lax spike. **Sepals** olive-green to pale green and more or less suffused with brownish violet, the lateral ones usually distinctly bicoloured (pale green/purplish brown) with the mid-vein constituting a boundary, 9-19 × 4-9 mm. **Petals** yellowish green to olive-green or dull purplish brown (occasionally suffused with rust-red), 5-13 × 1.5-4 mm, approximately half as wide as the sepals. **Lip** with reddish brown to blackish brown ground colour, now and then with a yellow margin, entire (rarely slightly three-lobed) with rounded to truncate base, 9-18(-20) × 9-20(-22) mm, approximately as long as the dorsal sepal, the basal part shaggy to velvety along the margin, the distal part velvety to subglabrous along the margin; front edge rounded to acuminate with a short terminal point; bulges obliquely conical (each of them at least half as large as the stigmatic cavity); **mirror** H-shaped or consisting of two parallel longitudinal bands. Eye-like knobs of the column dark (rarely pale). FIG. 85.

FIG. 86. FIG. 86. *Ophrys sphegodes* subsp. *cretensis* (Greece, Crete, Melambes, 24th March 1997). Photo by N. Faurholdt.

16.11. *Ophrys sphegodes* subsp. *cretensis*

Subsp. *cretensis* can be mistaken for the later flowering subsp. *gortynia* – but see the key. Flowering takes place from February to mid-April, with a peak in the second half of March.

Subsp. *cretensis* is endemic to the Greek islands, occurring mainly in Crete where it is locally common. It is also known from Karpathos and the Cycladean island of Paros and must be expected to occur on additional islands in the Aegean Sea. [Aeg, Cre].

DESCRIPTION. *Ophrys sphegodes* Mill. subsp. *cretensis* H. Baumann & Künkele Syn. *O. cretensis* (H. Baumann & Künkele) Paulus.

Plant relatively slender, 20-50 cm tall with 6-11 flowers in a lax spike. **Sepals** pale green (occasionally with a rose-coloured tinge), or the lateral ones distinctly bicoloured (pale green/pale purplish brown) with the mid-vein constituting a boundary, 8-11 × 4-5 mm. **Petals** yellowish green to olive-green (often suffused with rust-red), 5-9 × 1.5-2.5 mm, approximately half as wide as the sepals. **Lip** with brown ground colour, now and then provided with a narrow light brown to yellow margin, entire with rounded to truncate base, 5–9 × 7–10 mm, shorter than the dorsal sepal, the basal part shortly velvety to subglabrous, the distal part (sub)glabrous; front edge rounded to shortly acuminate with a short terminal point; bulges weakly developed to obliquely conical (but each of them at most one fourth as large as the stigmatic cavity); **mirror** H-shaped (rarely a little more complicated). Eye-like knobs of the column dark. FIG. 86.

16.12. *Ophrys sphegodes* subsp. *gortynia*

Subsp. *gortynia* can be confused with forms of subsp. *mammosa* and with the earlier flowering subsp. *cretensis*, but see the key. Flowering is relatively late, taking place from mid–April to early May.

Subsp. *gortynia* is endemic to the Greek islands where it has so far been found in Crete and the central Cyclades (Naxos, Paros and Antiparos). [Cre, Gre].

FIG. 87

FIG. 87. *Ophrys sphegodes* subsp. *gortynia* (Greece, Crete, Agia Triada, 13th April 2003). Photo by H. Æ. Pedersen.

DESCRIPTION. *Ophrys sphegodes* Mill. subsp. *gortynia* H. Baumann & Künkele Syn. *O. gortynia* (H. Baumann & Künkele) Paulus.

Plant relatively slender, 15-35 cm tall with 3-6 flowers in a lax spike. **Sepals** olive-green to pale green, rarely whitish, occasionally faintly suffused with violet towards the base, 8-11 × 4-6 mm. **Petals** pale green to yellowish green (often suffused with rust-red), 6-8 × 2.5-4 mm, approximately half as wide as the sepals. **Lip** with dark brown ground colour, occasionally provided with a narrow light brown to yellow margin, entire with wedge-shaped base, (8-)9-14 × 9-14 mm, usually longer than the dorsal sepal, the basal (and to less extent the distal) part velvety along the margin; front edge rounded with a short terminal point; bulges weakly developed to obliquely conical (but each of them at most half as large as the stigmatic cavity); **mirror** H-shaped or consisting of two parallel longitudinal bands. Eye-like knobs of the column dark. FIG. 87.

REFERENCES: Baumann & Künkele (1982, 1984, 1988), Bournérias (1998), Buttler (1986), E. Danesch & O. Danesch (1969), Davies et al. (1983), Del Prete & Tosi (1988), Delforge (1994, 2001, 2005), Hirth & Spaeth (1992), Kullenberg (1961, 1973), Lorella et al. (2002), Nelson (1962), Paulus (1988, 2000, 2001a), Paulus & Gack (1986, 1990b, 1995, 1999), Rossi (2002), Souche (2004), Sundermann (1980).

17. *Ophrys lunulata*

In Europe, many subspecies of bee orchids have narrow distributions. At species level, on the other hand, only *O. lunulata* is a true local endemic, being restricted to Sicily. Within its narrow range it often occurs abundantly, and with a little luck one can observe the pollination that is conducted by the megachilid bee *Osmia kohli*. *Ophrys lunulata* is characterised by the combination of long, narrow petals and a deeply three-lobed, apparently narrow lip with a usually crescent-shaped, centrally placed mirror. It varies little and can hardly be confused with other species (but possibly with forms of the partly stabilised hybrid complex *O. ×arachnitiformis*). Now and then flowers are seen in which the sides of the lip are not reflexed. This gives the lip a broader outline and exposes its yellowish to light brown marginal hairiness.

HABITAT. Dry to fairly moist, calcareous to slightly acid soil in full sunlight to light shade, from sea level to 800 m altitude. Typical habitats include meadows, grassland, garrigue and open woods.

FLOWERING TIME. March and April with a peak in the first half of April.
DISTRIBUTION. As mentioned above, *O. lunulata* appears endemic to Sicily,
where it is particularly frequent in the southeast and around Palermo.
Reports from Calabria, Elba, Malta, the Eolian Islands and Sardinia have not
been verified. The reports from Sardinia are probably due to confusion with
a form of *O. ×arachnitiformis* with reflexed lip sides. [Sic] MAP 19.
DESCRIPTION. *Ophrys lunulata* Parl.
Syn. *O. sphegodes* Mill. subsp. *lunulata* (Parl.) H. Sund.

Plant slender, 10-40 cm tall with (4-)6-10 flowers in a (relatively) lax spike.
Sepals pale rose-coloured to violet, rarely white, narrowly (oblong-)
lanceolate, 10-16 × 4.5-7 mm; dorsal sepal more or less boat-shaped,
incurved, from the base reflexed. **Petals** of approximately the same colour as
the sepals, linear-lanceolate with flat margins, 8-11 × 2-3 mm, glabrous or
velvety, spreading to slightly recurved. **Lip** with reddish brown to dark
brown ground colour and a yellowish to light brown margin, straight with

FIG. 88A

FIG. **88**. *Ophrys lunulata*
(**A,D.** Italy, Sicily, Ferla, 17th
April 1992; **B-C.** Italy, Sicily,
Ferla, 18th April 1999). Photos
by N. Faurholdt (**A,D**),
H. Æ. Pedersen (**B-C**).

Fig. 88c

Fig. 88b

Fig. 88d

Map 19. The total range of *Ophrys lunulata*. This species is not subdivided into subspecies.

reflexed (rarely spreading) sides, deeply three-lobed near the middle, 10–14 × 13–18 mm, shaggy (to velvety) on the side lobes and along the margin of the mid-lobe (otherwise subglabrous); bulges weakly developed (sometimes absent), distinctly isolated from the lip margin (side lobes flat); mid-lobe much longer than the side lobes, obtuse to rounded, provided with a short, more or less porrect (to downward pointing), triangular point; **mirror** distinct, consisting of a crescent- to horseshoe-shaped figure (less frequently of two drop-shaped spots), placed centrally without connection to the base of the lip (rarely connected to the base of the lip by fine, obscure lines), shining bluish grey to reddish brown. **Column** acute, not tapering towards the base (in lateral view); stigmatic cavity approximately as wide as long and approximately twice as wide as the anther, with lateral dark, eye-like knobs at base. Fig. 88.

References: Bartolo et al. (2001), Baumann & Künkele (1982, 1988), Buttler (1986), E. Danesch & O. Danesch (1969), Davies et al. (1983), Del Prete & Tosi (1988), Delforge (1994, 2001, 2005), Lorenz & Lorenz (2002), Nelson (1962), Paulus & Gack (1990c), Rossi (2002), Sundermann (1980).

Map 20. The European range of *Ophrys reinholdii*. Only subsp. *reinholdii* occurs in Europe.

18. *Ophrys reinholdii*

As in *O. kotschyi*, the flowers of *O. reinholdii* are remarkable for their sharp contrast between black and white on the lip. However, the two species are easily distinguished by details of the mirror. In *O. kotschyi*, the mirror consists of a large, H-shaped to considerably more complicated figure, the arms of which are connected to the base of the lip. In *O. reinholdii*, the mirror consists of one or two simple, centrally placed figures without connection to the base of the lip. *Ophrys reinholdii* is pollinated by one or two species of anthophorid bees belonging to the genus *Eupalovskia*.

The morphological variation mainly falls between two groups of populations which are also geographically separated. Based on this we recognise two subspecies of which only subsp. *reinholdii* is known from Europe (see, however, the section Dubious records p. 226 concerning subsp. *straussii*). Consequently, the present treatment deals with subsp. *reinholdii* only.

HABITAT. Dry to moist, calcareous to slightly acid soil in full sunlight to

FIG. 89. *Ophrys reinholdii* (**A.** Greece, Lesbos, Agiassos, 27th March 2002; **B.** Greece, Naxos, Koronos, 18th April 2000; **C:** Greece, Rhodes, Attaviros, 17th April 1996; **D.** Greece, Chios, Pelineo, 16th April 2005). Photos by A. Gøthgen (**A-B**), H. Æ. Pedersen (**C-D**).

partial shade, from sea level to 1300 m altitude. Typical habitats include roadside slopes, garrigue, old olive groves, graveyards and open to rather dense pine and oak woods.

FLOWERING TIME. From March to May, with a distinct peak in April on the Aegean Islands.

DISTRIBUTION. The southwestern part of the Balkan Peninsula, on the Ionian Islands and on the Aegean Islands (where it is common on Rhodes, for example). An isolated occurrence in southeastern Bulgaria was discovered as recently as 2003. The total range of the species also includes Cyprus, southern and southwestern Anatolia, Iran and Iraq. [Aeg, Alb, Gre, Yug; Ana, Cyp] MAP 20.

DESCRIPTION. *Ophrys reinholdii* Spruner ex H. Fleischm.

Plant slender, 15-50 cm tall with 2-10 flowers in a lax spike. **Sepals** pale rose-coloured to purplish violet (less frequently pale green to white), narrowly ovate to oblanceolate-oblong, 12-16 × 4-7.5 mm; dorsal sepal nearly flat, slightly incurved, from the base reflexed. **Petals** pale rose-coloured to green, often suffused with brown or red, narrowly triangular (occasionally auriculate) with flat margins, 4-8 × 1.5-2.5 mm, velvety to shaggy, spreading to somewhat reflexed. **Lip** with black to dark brown ground colour, straight with more or less recurved margin, deeply three-lobed at the base, 10-15 × 12-16 mm, shaggy (of frequently silvery hairs) on the side lobes, velvety along the margin of the mid-lobe (otherwise subglabrous); side lobes converted into obliquely conical (to weakly developed) bulges; mid-lobe much longer than the side lobes, rounded, provided with a short, more or less porrect, triangular point; **mirror** distinct, consisting of a more or less horseshoe-shaped figure, or two drop-shaped figures, without connection to the base of the lip, white or shining greyish violet with a white border. **Column** acute, not tapering towards the base (in side view); stigmatic cavity wider than long and approximately twice as wide as the anther, devoid of lateral, eye-like knobs at base. FIG. 89.

REFERENCES: Baumann & Künkele (1982, 1988), Bergman et al. (2004), Buttler (1986), Davies et al. (1983), Delforge (1994, 2001, 2005), Nelson (1962), Paulus (2001b), Paulus & Gack (1986, 1990b), Sundermann (1980).

FIG. 89C

FIG. 89D

19. *Ophrys kotschyi*

The flowers of O. *kotschyi* are remarkable for the contrasting play of black and white on the lip. This species has a superficial similarity to O. *reinholdii*, but it is easily distinguished by its H-shaped to far more complicated mirror, which is connected to the base of the lip by two broad bands. Geographically, the two species seem to overlap only in mainland Greece (where O. *kotschyi* is very rare), on a few Aegean Islands (e.g. Naxos) and in Cyprus.

In Europe, the morphological variation mainly falls between two groups of populations. These entities appear to be maintained through pollinator specificity, for which reason we recognise them as separate subspecies. *Ophrys kotschyi* is pollinated by anthophorid bees of the genus *Melecta*. Subsp. *cretica* seems to be pollinated exclusively by *M. tuberculata*, whereas subsp. *ariadnae* is pollinated by *M. albifrons*.

HABITAT. Calcareous, dry to moist soil in full sunlight to light shade, from sea level to 1000 m altitude. Typical habitats include roadside slopes, grassland, garrigue and old, pesticide-free olive groves.

FLOWERING TIME. From mid-February to early May, with a peak from mid-March to mid-April.

DISTRIBUTION. The geographic range of O. *kotschyi* encompasses the southeastern part of mainland Greece, a number of islands in the southern part of the Aegean Sea (north to Chios) and Cyprus. In the latter island it is represented by the endemic subsp. *kotschyi* only. [Aeg, Cre, Gre; Cyp] MAP 21.

DESCRIPTION. *Ophrys kotschyi* H. Fleischm. & Soó

Synonyms are listed under the subspecies.

Plant compact to slender, 10-40 cm tall with 2-10 flowers in a relatively dense to lax spike. **Sepals** green to brownish (the lateral ones sometimes bicoloured with the mid-vein constituting a boundary), rarely rose-coloured, narrowly ovate to (ob)lanceolate, (8-)10-17 × 5-8.5 mm; dorsal sepal more or less boat-shaped to nearly flat, slightly incurved, from the base reflexed. **Petals** green to reddish brown or rust-red, triangular-lanceolate (occasionally auriculate) with flat margins, (4-)6-9 × 1.5-3 mm, velvety, recurved to spreading. **Lip** with dark brown to purplish black ground colour, straight with recurved margins, shallowly to deeply three-lobed near the base, 10-16 × 10-16 mm, velvety to shaggy on the side lobes, velvety along the margin of the mid-lobe (otherwise subglabrous); side lobes converted into weakly developed to obliquely conical bulges; mid-lobe distinctly longer than the side lobes, obtuse to truncate, provided with a short, more or less porrect, triangular point; **mirror** distinct, consisting of a large, H-shaped to far more complicated figure, the basal arms of which are connected to the base of the lip, dull greyish violet to reddish brown with a white border. **Column** acute (to obtuse), not tapering towards the base (in side view); stigmatic cavity approximately as wide as long and approximately twice as wide as the anther, devoid of lateral, eye-like knobs at its base. FIGS 90-91.

Key to the subspecies

1. Outline of the stigmatic cavity transversely elliptic to nearly circular. Side lobes of the lip often converted into pronounced, obliquely conical bulges (less frequently just slightly vaulted) . . . . 19.1. **subsp.** *cretica*
1. Outline of the stigmatic cavity rectangular to nearly square. Side lobes of the lip slightly vaulted . . . . 19.2. **subsp.** *ariadnae*

MAP 21. The total range of *Ophrys kotschyi*. Two subspecies are recognized in Europe; their distributions overlap in Crete, Paros and Naxos.

FIG. 90A

FIG. 90B

FIG. 90. *Ophrys kotschyi* subsp. *cretica* (**A.** Greece, Crete, Agia Triada, 13th April 2003; **B.** Greece, Naxos, Halkio, 16th April 2000). Photos by A. Gøthgen (**B**), H. Æ. Pedersen (**A**).

19.1. *Ophrys kotschyi* subsp. *cretica*

Subsp. *cretica* can be mistaken for the generally slightly earlier flowering subsp. *ariadnae* – but see the key. In central and western Crete, small-flowered, late-flowering individuals with a greatly reduced mirror are occasionally encountered. On Naxos, subsp. *cretica* is known from low, calcareous hills to the west, between the villages of Galanado and Ano Sangri. Here, as well as on southern Rhodes, a large proportion of the plants have rose-coloured sepals and rust-red petals. Subsp. *cretica* generally flowers a little later than subsp. *ariadnae*, from late March to early May.

Subsp. *cretica* is known with certainty from the Greek province of Attika, the Peloponnese and Aigina (everywhere very rare), as well as from Crete, Paros, Psara and Inouses near Chios, Naxos (very local) and southern Rhodes. [Aeg, Cre, Gre].

DESCRIPTION. *Ophrys kotschyi* H. Fleischm. & Soó subsp. *cretica* (Vierh.) H. Sund.
Syn. *O. cretica* (Vierh.) E. Nelson subsp. *beloniae* G. & H. Kretschmar; *O. cretica* (Vierh.) E. Nelson subsp. *bicornuta* H. Kretzschmar & R. Jahn; *O. cretica* (Vierh.) E. Nelson s.s.; *O. doerfleri* auct. [non H. Fleischm.?].

Plant 10-40 cm tall with 2-10 flowers in a relatively dense (to lax) spike.
Sepals green to brownish violet (rarely rose-coloured; the lateral ones rarely bicoloured with the mid-vein constituting a boundary), (8-)10-15 × 5-7.5

mm. **Lip** 10-14 × 10-14 mm; side lobes often converted into pronounced, obliquely conical bulges (less frequently just slightly vaulted); mid-lobe 9-13.5 × 10-16.5 mm, with a 1-2 mm long terminal point. Stigmatic cavity transversely elliptic to nearly circular in outline. FIG. 90.

19.2. *Ophrys kotschyi* subsp. *ariadnae*

This subspecies can be mistaken for the generally slightly later flowering subsp. *cretica*, but see the key. Flowering is relatively early, from late February to mid-April.

Subsp. *ariadnae* is so far known from central and western Crete (locally common), Karpathos, Naxos (common), Paros and Psara west of Chios. [Cre, Gre].

FIG. 91A

FIG. 91. *Ophrys kotschyi* subsp. *ariadnae* (**A.** Greece, Crete, Kissou Kambos, 12th April 2003; **B-C.** Greece, Naxos, Galnado, 16th April 2000). Photos by H. Æ. Pedersen (**A**), N. Faurholdt (**B**), A. Gøthgen (**C**).

DESCRIPTION. *Ophrys kotschyi* H. Fleischm. & Soó subsp. *ariadnae* (Paulus) N. Faurholdt

Syn. *O. ariadnae* Paulus; *O. cretica* (Vierh.) E. Nelson subsp. *ariadnae* (Paulus) H. Kretzschmar; *O. cretica* (Vierh.) E. Nelson subsp. *karpathensis* E. Nelson, nom. inval.; *O. cretica* (Vierh.) E. Nelson subsp. *naxia* E. Nelson, nom. inval.

Plant 10-30 cm tall with 2-8 flowers in a lax (to relatively dense) spike. **Sepals** green or the lateral ones bicoloured (green/brownish violet with the mid-vein constituting a boundary), (10-)12-17 × 6-8.5 mm. **Lip** 12-16 × 12-16 mm; side lobes slightly vaulted; mid-lobe 10.5-16 × 12.5-18.5 mm, with a 1-3 mm long terminal point. Stigmatic cavity rectangular to nearly square in outline. FIG. 91.

REFERENCES: Baumann & Künkele (1982, 1988), Buttler (1986), E. Danesch & O. Danesch (1969), Davies et al. (1983), Delforge (1994, 2001, 2005), Gölz & Reinhard (1996), Nelson (1962), Paulus (1988, 1994, 2001b), Paulus & Gack (1986, 1990b, 1992), Saliaris (2002), Sundermann (1980), Taylor (2005), Vöth (1986).

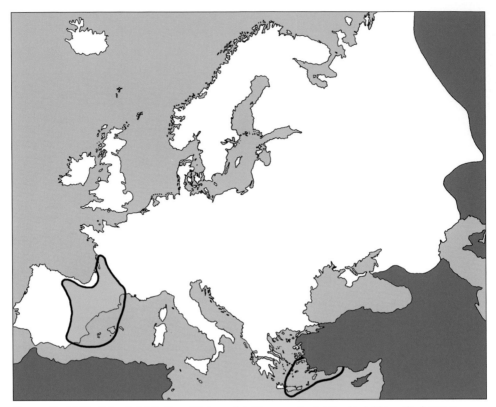

MAP 22. The highly disjunct distribution of *Ophrys* ×*brigittae*, a partly stabilised hybrid complex between *O. fusca* and *O. omegaifera*.

H1. *Ophrys* ×*brigittae*

In both the eastern and the western part of South Europe, *Ophrys* populations occur which are morphologically intermediate between *O. fusca* and *O. omegaifera*, and which are commonly accepted as two distinct species: the western *O. vasconica* and the eastern *O. sitiaca*. We, however, consider these populations a partly stabilised hybrid complex (evidently with different subspecies of *O. omegaifera* involved in the eastern and western parts of the range). For this reason we have chosen to treat them under the oldest validly published name for the hybrid, *O. fusca* × *omegaifera*: *O.* ×*brigittae*. In the eastern part of its range, *O.* ×*brigittae* is reported to be pollinated by the andrenid bee *Andrena nigroaena*, a pollinator that it shares with certain forms of *O. fusca* subsp. *fusca*.

HABITAT. Dry to humid soil in full sunlight to light shade. Typical habitats include garrigue and open pine woods.

FIG. 92A

FIG. 92B

FIG. 92. *Ophrys ×brigittae* (**A.** Fra, Languedoc, Bugarac, 7th May 2002; **B.** Greece, Samos, Agia Kyriaki, 11th April 1999). Photos by C. A. J. Kreutz (**A**), H. Æ. Pedersen (**B**).

FLOWERING TIME. From March to June in the western part of its range and from January to April in the eastern part.

DISTRIBUTION. The distribution is disjunct, encompassing a western Mediterranean area (Mallorca, northeastern Spain, southwestern France) and an eastern Mediterranean area (the islands in the southeastern part of the Aegean Sea, southwestern Anatolia). Reports from Portugal should be checked. [Aeg, Bal, Cre, Fra, Spa; Ana] MAP 22.

DESCRIPTION. *Ophrys ×brigittae* H. Baumann

Syn. *O. sitiaca* Paulus et al.; *O. vasconica* (O. & E. Danesch) P. Delforge; *O. fusca* Link subsp. *vasconica* O. & E. Danesch; *O. fusca* Link var. *vasconica* (O. & E. Danesch) Landwehr, comb. inval.

Plant compact to slender, 7-20(-35) cm tall with 1-6(-12) flowers in a lax to relatively dense spike. **Sepals** (pale) green, ovate to elliptic, 8-16 × 5.5-10 mm; dorsal sepal boat-shaped, incurved, from the base nearly parallel to the column. **Petals** (yellowish) green, often suffused with brown (especially along the margin), narrowly oblong with (nearly) flat margins, 6-11 × 1.5-4 mm, glabrous, spreading to moderately incurved. **Lip** with brown to grey or nearly black ground colour and sometimes a narrow yellow margin, straight to gradually downcurved with recurved margins, obscurely longitudinally furrowed at base, three-lobed at the middle or closer to the apex, (11-)13-16(-19) × 8-15.5 mm, velvety to shaggy (particularly towards the margin); bulges absent; mid-lobe distinctly longer than the side lobes, emarginate (to rounded); **mirror** distinct (to somewhat obscure), consisting of a figure shaped more or less like the tail of a fish, stretching from the base to above the middle of the lip, dull (reddish) brown to greyish violet (often strongly marbled), in front delimited by a conspicuous white to bluish grey, ω-shaped band and occasionally with a pale area at base. **Column** rounded, not tapering towards the base (in side view); stigmatic cavity usually wider than long and approximately twice as wide as the anther, devoid of lateral, eye-like knobs at base. Fig. 92.

REFERENCES: Baumann & Künkele (1988), Bournérias (1998), E. Danesch & O. Danesch (1969), Delforge (1994, 2001, 2005), Paulus (1988, 2001b), Paulus & Gack (1990b), Souche (2004).

H2. *Ophrys* ×*vicina*

Ophrys heterochila was originally described as a subspecies of the present *O. fuciflora*. However, as particularly demonstrated by the variation in the incisions of the lip and the shape and size of the bulges, this entity represents a gradual transition between forms of *O. fuciflora* on the one side and forms of *O. scolopax* on the other (FIG. 25). Therefore, we believe that it is in fact a partly stabilised hybrid complex between *O. fuciflora* subsp. *fuciflora* and *O. scolopax* s.l. For similar reasons, we consider *O. calypsus* (including the so-called *O. heldreichii* var. *pseudoapulica* and var. *scolopaxoides*) a partly stabilised hybrid complex between *O. fuciflora* subsp. *fuciflora* and *O. scolopax* subsp. *heldreichii*, just as we think that *O. holubyana* (including its later synonym, *O. zinsmeisteri*) is a partly stabilised hybrid complex between *O. fuciflora* subsp. *fuciflora* and *O. scolopax* subsp. *cornuta*. The so-called *O. heterochila*, *O. calypsus* and *O. holubyana* thus share their putative parents (at species level), and we have consequently chosen to combine the three latter entities under the oldest validly published name for the hybrid between *O. fuciflora* s.l. and *O. scolopax* s.l.: *O.* ×*vicina*.

The so-called *O. calypsus*, *O. heterochila* and *O. holubyana* are not difficult to tell apart (in fact, they are markedly different), but as their

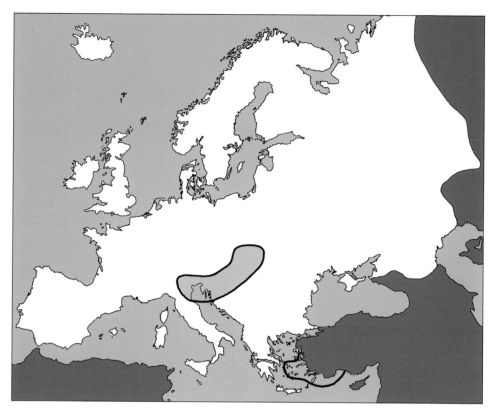

Map 23. The disjunct European distribution of *Ophrys ×vicina*, a partly stabilised hybrid complex between *O. fuciflora* and *O. scolopax*. The nothomorph *"holubyana"* is confined to the northern range, whereas the nothomorphs *"calypsus"* and *"heterochila"* overlap in the southern range.

parents have only been partly identified so far, we have chosen not to recognise them formally as separate entities. For the time being, it seems most appropriate simply to refer to them as informal hybrid forms, so-called "nothomorphs" (nm.).

Ophrys ×vicina nm. *"calypsus"* (Fig. 93A–C) is 15–40(–50) cm tall. It is characterised by its large flowers (petals (5–)6–7(–8.5) mm long, lip 12–18 mm wide), which have spreading to slightly reflexed sepals and a lip that is markedly vaulted (in side view) and provided with 2.5–5 mm long, straight (to slightly falcate) bulges. Individual plants often carry both *O. fuciflora*-like flowers with entire lips and *O. scolopax*-like flowers with three-lobed lips.

Ophrys ×vicina nm. *"heterochila"* (Fig. 93D–G) is (7,5)9–20(–27) cm tall. It is characterised by its small flowers (petals 2.5–3 mm long, lip 8–11 mm wide), which have strongly reflexed sepals that almost cover the ovary and a lip that is only slightly vaulted (in side view) and provided with 1–2.5 mm long, straight to distinctly falcate bulges. The large diameter of the basal rosette of leaves (as compared with the height of the plant) is likewise

FIG. 93A

FIG. 93B

FIG. 93C

FIG. 93D

FIG. 93E

FIG. 93. *Ophrys ×vicina*. **A-C.** nm. "*calypsus*"; **D-G.**
nm. "*heterochila*"; **H-I.** nm. "*holubyana*" (**A-B.**
Greece, Rhodes, Filerimos, 14th April 1996; **C.**
Greece, Naxos, Halkio, 16th April 2000; **D.** Greece,
Chios, Pelineo, 16th April 2005; **E-F.** Greece,
Rhodes, Profitis Ilias, 15th April 1996; **G.** Greece,
Rhodes, Kalathos, 25th March 2002; **H-I.** Czech
Republic, Bilé Karpaty, 29th May 1994).
Photos by H. Æ. Pedersen (**A-B**, **D-E**), N. Faurholdt
(**C**, **F-G**), C. A. J. Kreutz (**H-I**).

FIG. 93G

FIG. 93F

FIG. 93H

FIG. 93I

characteristic of nm. "*heterochila*". Individuals with *O. scolopax*-like flowers from the large, diverse occurrence of these putative hybrids on the Rhodean mountain of Profitis Ilias were recently described as a new (in our opinion dubious) species, *O. dodekanensis*.

Ophrys ×*vicina* nm. "*holubyana*" (FIG. 93H-I) is characterised by its medium-sized flowers (petals 3-6 mm long, lip 10-17 mm wide), which have spreading to slightly reflexed sepals and a lip that is fairly vaulted (in side view) and provided with 2-8 mm long, distinctly falcate bulges.

The pollination of *O.* ×*vicina* is poorly known; nm. "*heterochila*" and nm. "*holubyana*" are reported to be pollinated by the anthophorid bees *Eucera cypria* and *E. longicornis*, respectively.

HABITAT. Dry to moist soil in full sunlight to light shade in garrigue, old olive groves and light pine woods, as well as in open places in maquis, up to 700 m altitude.

FLOWERING TIME. Flowering in nm. "*calypsus*" and nm. "*heterochila*" takes place early, mainly in March and the first half of April (although flowering individuals may be encountered from late February to early May). Nm. "*holubyana*" flowers in May and June.

DISTRIBUTION. East Mediterranean and Southeast European. Nm. "*calypsus*" is known from a series of islands in the Aegean Sea (from Siros in the northwest to Rhodes in the southeast) and from adjoining parts of southwestern Anatolia. Nm. "*heterochila*" occurs in southwestern Anatolia and on the Sporadhes (from Samos in the north to Rhodes in the south). Reports from the region of Koronis on Naxos are probably due to confusion with fortuitous hybrids between *O. fuciflora* subsp. *andria* and *O. scolopax* subsp. *scolopax*. Nm. "*holubyana*" is distributed from former Yugoslavia north to the Carpathian Mountains and west to northeasternmost Italy. Also, the so-called *O. homeri*, mainly distributed on the eastern Aegean islands, arguably belongs to *O.* ×*vicina*, but it cannot be referred unequivocally to one of the above nothomorphs. [Aeg, Cze, Gre, Hun, Ita, Yug; Ana] MAP 23.

DESCRIPTION. *Ophrys* ×*vicina* Duffort
Syn. *O. calypsus* M. Hirth & H. Spaeth; *O. heldreichii* Schltr. var. *calypsus* (M. Hirth & H. Spaeth) P. Delforge; *O. dodekanensis* H. Kretzschmar & Kreutz; *O. heterochila* (Renz & Taubenheim) P. Delforge; *O. holoserica* (Burm.f.) Greuter subsp. *heterochila* Renz & Taubenheim; *O. holubyana* András.; *O. homeri* M. Hirth & H. Spaeth; *O. heldreichii* Schltr. var. *pseudoapulica* P. Delforge; *O. heldreichii* Schltr. var. *scolopaxoides* P. Delforge; *O. zinsmeisteri* A. Fuchs & Ziegensp.

Plant compact to slender, (7.5-)9-40(-50) cm tall with (1-)2-10 flowers in a dense to lax spike. **Sepals** (greenish) white to violet, (ob)ovate to elliptic, 9-15(-17.5) × 4-7(-11) mm; dorsal sepal shallowly to moderately boat-shaped, more or less incurved, from the base reflexed. **Petals** of almost the same colour as the sepals, triangular to oblong (often auriculate) with recurved margins, 2.5-7(-8.5) × 1-3.5 mm, shaggy, spreading. **Lip** with (reddish) brown to dark brown ground colour, straight with more or less recurved (rarely spreading) sides, entire to deeply (and often asymmetrically) three-lobed close to the base, 7-13.5(-15) × 8-20 mm, along the margin velvety towards the apex and brownish-shaggy towards the base (also on the outer side of the side lobes, if present), otherwise (sub)glabrous; bulges very variable, weakly developed to obliquely conical or horn-like, in some cases isolated from the margin of the lip, in other cases it is the side lobes that are converted into bulges; mid-lobe longer than the side lobes (if present), rounded (to obtuse), provided with a usually broad and conspicuous, erect (to porrect), rectangular, rhombic or (ob)triangular appendage; **mirror** distinct, consisting of an H- or X-shaped to considerably more complicated figure, the basal arms of which are usually connected to the base of the lip, dull greyish blue to violet (rarely reddish brown) with an often broad cream border. **Column** acute to rounded, not tapering towards the base (in side view); stigmatic cavity at least as wide as long and approximately twice as wide as the anther, with lateral, dark, eye-like knobs at base. FIG. 93.

REFERENCES: Baumann & Künkele (1988), Delforge (1994, 2001, 2005), Gulyás et al. (2005), Molnár & Gulyás (2005), Paulus (2001b), Paulus & Gack (1990b, 1992), Pedersen & Faurholdt (1997), Vöth & Ehrendorfer (1976).

H3. *Ophrys ×delphinensis*

Ophrys ×delphinensis is believed to have originated as a hybrid between *O. argolica* subsp. *argolica* and *O. scolopax* (probably subsp. *cornuta*), and experimentally produced first-generation plants of *O. argolica* subsp. *argolica* × *scolopax* subsp. *cornuta* do look very similar to *O. ×delphinensis* (Svante Malmgren, in litt.). This entity is recognised by many authors as a distinct species, *O. delphinensis*. However, due to its marked variation that particularly complicates delimitation towards *O. argolica*, we prefer to treat *O. ×delphinensis* as a partly stabilised hybrid complex. The flowers are pollinated by the anthophorid bee *Anthophora plagiata*, a species that also pollinates *O. argolica* subsp. *argolica*.

MAP 24. The total range of *Ophrys ×delphinensis*, a partly stabilised hybrid complex between *O. argolica* and *O. scolopax*.

HABITAT. Calcareous, dry to moist soil in full sunlight to light shade, up to 1100 m altitude. Typical habitats include garrigue, open woods, old pesticide-free olive groves, grassland and seepage meadows.

FLOWERING TIME. April and May.

DISTRIBUTION. This hybrid complex is only known from Greece, where it occurs in Euboea and adjoining parts of the mainland, south to the Gulf of Korinthos. [Gre] MAP 24.

DESCRIPTION. *Ophrys ×delphinensis* O. & E. Danesch.

Plant slender, 20-40 cm tall with 3-10 flowers in a (relatively) lax spike. **Sepals** (purplish) violet to bright rose-coloured, ovate to elliptic, 12-16 × 6-9 mm; dorsal sepal shallowly boat-shaped to nearly flat, slightly incurved, from the base reflexed. **Petals** of approximately the same colour as the sepals, (triangular-)oblong with (nearly) flat margins, 4.5-7 × 2-4 mm, shaggy, spreading. **Lip** with reddish brown to dark brown ground colour, straight with strongly recurved sides, deeply three-lobed close to the base,

FIG. 94. *Ophrys* ×*delphinensis* (**A-B.** Greece, the Peleponnese, Diakofto 9th May 1997). Photos by A. Gøthgen.

10-12.5 × 9-16 mm, densely white-shaggy on the outer side of the side lobes and velvety along the margin of the mid-lobe (otherwise subglabrous); side lobes converted into weakly developed to obliquely conical bulges; mid-lobe distinctly longer than the side lobes, obtuse to rounded, provided with a broad and conspicuous, more or less porrect, rectangular, rhombic, or (ob)triangular appendage; **mirror** distinct, often consisting of an H-shaped to horseshoe-shaped (but very variable) figure placed centrally with or without connection to the base of the lip, dull (to shining) greyish blue. **Column** acute to obtuse, not tapering towards the base (in side view); stigmatic cavity at least as wide as long and approximately twice as wide as the anther, with dark or pale, lateral, eye-like knobs at base. FIG. 94.

REFERENCES: Baumann & Künkele (1982, 1988), Buttler (1986), O. Danesch & E. Danesch (1976), Davies et al. (1983), Delforge (1994, 2001, 2005), Paulus & Gack (1990b).

H4. *Ophrys ×arachnitiformis*

Ophrys ×arachnitiformis was originally attributed status as a species. Attempts have since been made to divide it into a range of more narrowly circumscribed species which, however, are poorly delimited towards each other. Furthermore, it is difficult to draw a boundary between *O. fuciflora* on the one hand and, for example, the so-called *O. tyrrhena* on the other – or between *O. sphegodes* on the one hand and, for example, the so-called *O. splendida* on the other. Because of these difficulties, we have decided to treat *O. arachnitiformis* s.s., *O. archipelagi*, *O. aveyronensis*, *O. exaltata* subsp. *marzuola*, *O. massiliensis*(?), *O. morisii*, *O. panattensis*, *O. splendida* and *O. tyrrhena* as one (morphologically and phenologically highly variable) entity and to consider this a partly stabilised hybrid complex between *O. fuciflora* and *O. sphegodes*. At least part of the complex seems to have originated from the hybrid *O. fuciflora* subsp. *fuciflora* × *sphegodes* subsp. *sphegodes*, but in the total complex there are probably more subspecies involved on either side. On top of this, it cannot be ruled out that there are more *species* involved;

MAP 25. The total range of *Ophrys ×arachnitiformis*, a partly stabilised hybrid complex between *O. fuciflora* and *O. sphegodes*.

for example, it has been proposed that the so-called *O. morisii* is related genetically to *O. argolica* subsp. *crabronifera*. In Sardinia and southern Corsica, individuals are frequently encountered that are reminiscent of the latter with regard to the ground colour of the lip and the shape of the mirror. Perhaps, subsp. *crabronifera* has formerly occurred in those islands but disappeared through hybridisation and back-crossing?

Like *O. sphegodes*, *O. ×arachnitiformis* is predominantly pollinated by andrenid bees of the genus *Andrena* (but almost consistently other species). Additional pollinators of *O. ×arachnitiformis* include the colletid bee *Colletes cunicularius*, the anthophorid bee *Anthophora sicheli*, and a couple of megachilid bee species of the genus *Osmia*.

HABITAT. The ecological range is wide. *O. ×arachnitiformis* grows in calcareous to acid, dry to moist soil in full sunlight to light shade (rarely heavy shade), up to 1000 m altitude. Typical habitats include garrigue, maquis, grassland, roadside slopes, wooded meadows and open woods.
FLOWERING TIME. The various components of the hybrid complex have only partly overlapping flowering times. The total flowering time is therefore protracted, lasting from March to early June. Occasionally, a few flowering individuals are encountered as early as January or February.
DISTRIBUTION. The individual components of this hybrid complex have fairly local, but partly overlapping, distributions. The total range extends from northern Spain across southern France, Corsica and Sardinia to the western and southern parts of mainland Italy and to islands along the Dalmatian coast. [Cor, Fra, Ita, Sar, Yug] MAP 25.
DESCRIPTION. *Ophrys ×arachnitiformis* Gren. & Philippe [pro sp.] Syn. *O. exaltata* Ten. subsp. *arachnitiformis* (Gren. & Philippe) Del Prete; *O. sphegodes* Mill. subsp. *arachnitiformis* (Gren. & Philippe) H. Sund.; *O. archipelagi* Gölz & H. R. Reinhard; *O. exaltata* Ten. subsp. *archipelagi* (Gölz & H. R. Reinhard) Del Prete; *O. aveyronensis* (J. J. Wood) P. Delforge; *O. castellana* J. Devillers-Terschuren & P. Devillers; *O. sphegodes* Mill. subsp. *integra* (Moggr. & Rchb.f.) H. Baumann & Künkele; *O. exaltata* Ten. subsp. *marzuola* Geniez et al.; ?*O. massiliensis* Viglione & Véla; *O. montis-leonis* O. & E. Danesch; *O. morisii* (Martelli) Soó; *O. exaltata* Ten. subsp. *morisii* (Martelli) Del Prete; *O. panattensis* Scrugli et al.; *O. splendida* Gölz & H. R. Reinhard; *O. exaltata* Ten. subsp. *splendida* (Gölz & H. R. Reinhard) Soca; *O. tyrrhena* Gölz & H. R. Reinhard; *O. exaltata* Ten. subsp. *tyrrhena* (Gölz & H. R. Reinhard) Del Prete.

Plant slender, 10-40(-55) cm tall with 2-10(-12) flowers in a lax to relatively dense spike. **Sepals** white to rose-coloured, less frequently (purplish) violet or pale green, ovate to elliptic or (ob)lanceolate-oblong, 10-17 × (3-)4-9 mm; dorsal sepal nearly flat to shallowly boat-shaped, slightly incurved, from the base reflexed. **Petals** (whitish) yellow to rose-coloured or green, often suffused with red, (triangular-)oblong to elliptic or linear–lanceolate (sometimes auriculate) with flat to wavy or somewhat recurved margins, 5-11.5 × 2-5.5 mm, glabrous to velvety, recurved to spreading. **Lip** with dark brown to reddish brown ground colour and sometimes a narrow yellow margin, straight with more or less recurved (rarely spreading) sides, entire (to moderately three-lobed near the middle), 8-16(-18) × 8-22 mm, velvety to shaggy along the margin (otherwise subglabrous); bulges weakly developed (occasionally missing), distinctly isolated from the lip margin (side lobes flat, if present); front edge rounded (to obtuse), normally provided with a short, more or less porrect, triangular point; **mirror** distinct (rarely obscure), consisting of an H-shaped figure or of a simpler or slightly more complicated figure derived from this, distinctly connected to the base of the lip, dull (to shining) greyish blue to violet (rarely reddish brown), occasionally marbled and/or provided with a pale border. **Column** acute (to obtuse), not tapering towards the base (in side view); stigmatic cavity approximately as wide as long and approximately twice as wide as the anther, with dark or pale, lateral, eye-like knobs at base. FIG. 95.

REFERENCES: Baumann & Künkele (1982, 1988), Berger (2003), Bournérias (1998), Buttler (1986), E. Danesch & O. Danesch (1969), Davies et al. (1983), Del Prete & Tosi (1988), Delforge (1994, 2001, 2005), Gölz & Reinhard (1980), Hermosilla & Soca (1999), Hertel & Hertel (2003), Nelson (1962), Paulus & Gack (1990c, 1995, 1999), Rossi (2002), Souche (2004), Sundermann (1980).

FIG. 95A

FIG. 95B

FIG. 95C

FIG. 95. *Ophrys ×arachnitiformis* (**A-B.** Italy, Sardinia, Osini, 20th April 1997, **C.** Italy, Sardinia, Ponte Soala Manna, 22nd April 1997; **D.** Italy, Sardinia, Dorgali, 22nd April 1997; **E.** France, Provence, La Madeleine, 30th April 1990; **F.** Italy, Toscana, Castagneto, 12th April 2001). Photos by N. Faurholdt (**B,D,F**), H. Æ. Pedersen (**A,C,E**).

FIG. 95E

FIG. 95D

FIG. 95F

FIG. 96A

FIG. 96. *Ophrys ×flavicans* (**A-B.** France, Provence, Drap, 5th May 1990; **C.** Spain, Balearic Islands, Mallorca, Galilea, 18th April 1990; **D-E.** Italy, Puglia, Monte Gargano, 11th April 1995; **F.** Italy, Puglia, Mottola, 20th April 2002). Photos by H. Æ. Pedersen (**A-B**), N. Faurholdt (**C-F**).

H5. *Ophrys* ×*flavicans*

Ophrys ×*flavicans* was originally described as a narrowly defined species from former Yugoslavia. The synonyms below designate later described, likewise narrowly defined, entities from other parts of the central and western Mediterranean. All of these entities are poorly delimited from each other. Furthermore, it is difficult to draw a boundary between *O. sphegodes* on the one hand and, for example, the so-called *O. tarentina* on the other – or between *O. bertolonii* on the one hand and the so-called *O. aurelia* on the other. Because of these difficulties, we have decided to treat the "species" *O. aurelia*, *O. balearica*, *O. benacensis*, *O. bertoloniiformis*, *O. catalaunica*, *O. drumana*, *O. explanata*, *O. flavicans* s.s., *O. magniflora*, *O. melitensis*, *O. promontorii*, *O. saratoi* and *O. tarentina* as one (morphologically and phenologically highly variable) entity and to consider this a partly stabilised hybrid complex between *O. bertolonii* and *O. sphegodes* (evidently with more than one subspecies of the latter involved).

Different local forms of *O.* ×*flavicans* are pollinated by different species

FIG. 96B

FIG. 96C

FIG. 96D

FIG. 96E

FIG. 96F

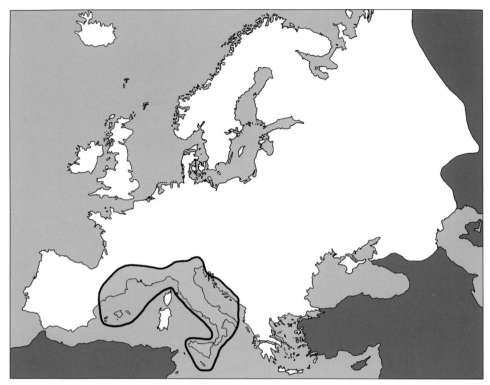

MAP 26. The total range of *Ophrys* ×*flavicans*, a partly stabilised hybrid complex between *O. bertolonii* and *O. sphegodes*.

of megachilid bees – mainly of the genus *Chalicodoma* (an adaptation shared with *O. bertolonii*), but also of the genus *Osmia*.

HABITAT. The ecological range of this complex is wide. *O.* ×*flavicans* grows in dry to moist soil in full sunlight to light shade, up to 1300 m altitude. Typical habitats include garrigue, open maquis, olive groves, grassland and open patches in forest; now and then it even occurs in waste ground.
FLOWERING TIME. The various components of the hybrid complex have only partly overlapping flowering times. The total flowering season is from March to June.
DISTRIBUTION. The individual components of this complex have fairly local, but partly overlapping, distributions. The total range extends from Cataluña across the Balearic Islands, southern France and mainland Italy to Dalmatia, Sicily and Malta. [Bal, Fra, Ita, Mal, Sic, Spa, Yug] MAP 26.
DESCRIPTION. *Ophrys* ×*flavicans* Vis. [pro sp.]
Syn. *O. aurelia* P. Delforge et al.; *O. balearica* P. Delforge; *O. benacensis* (Reisigl) O. Danesch et al.; *O. bertoloniiformis* O. & E. Danesch subsp. *benacensis* Reisigl; *O. saratoi* E. G. Camus subsp. *benacensis* (Reisigl) Del

Prete; *O. bertoloniiformis* O. & E. Danesch s.s.; *O. bertolonii* Moretti subsp. *bertoloniiformis* (O. & E. Danesch) H. Sund.; *O. pseudobertolonii* Murr subsp. *bertoloniiformis* (O. & E. Danesch) H. Baumann & Künkele; *O. catalaunica* O. & E. Danesch; *O. bertolonii* Moretti subsp. *catalaunica* (O. & E. Danesch) Soca; *O. pseudobertolonii* Murr subsp. *catalaunica* (O. & E. Danesch) H. Baumann & Künkele; *O. drumana* P. Delforge; *O. explanata* (Lojac.) P. Delforge; *O. magniflora* Geniez & Melki; *O. bertolonii* Moretti subsp. *magniflora* (Geniez & Melki) Soca; *O. melitensis* (Salkowski) J. Devillers-Terschuren & P. Devillers; *O. promontorii* O. & E. Danesch; *O. pseudobertolonii* Murr s.s.; *O. saratoi* E. G. Camus s.s.; *O. bertolonii* Moretti subsp. *saratoi* (E. G. Camus) Soca; *O. tarentina* Gölz & H. R. Reinhard.

Plant slender to relatively compact, 9-35(-48) cm tall with (1-)2-10(-13) flowers in a lax spike. **Sepals** (yellowish) green to white or violet (the lateral ones rarely bicoloured with the mid-vein constituting a boundary), narrowly elliptic to (ob)lanceolate-oblong, 8-17 × 2-9 mm; dorsal sepal (nearly) flat, slightly incurved, from the base reflexed. **Petals** rose-coloured to violet or yellowish green, often suffused with red or brown, oblong to lanceolate-triangular with flat or wavy margins, 5.5-12 × 2-6 mm, glabrous to pubescent, spreading to slightly incurved or recurved. **Lip** with light reddish brown to blackish brown ground colour, moderately saddle-shaped to straight and with more or less recurved sides, entire (to moderately three-lobed near the middle), 8-21 × 10-17 mm, shaggy along the margin, more or less velvety on the central part (though not on the mirror); bulges absent or only weakly developed, distinctly isolated from the lip margin (side lobes flat, if present); front edge rounded to emarginate, provided with a short, erect (rarely porrect), narrowly triangular (to awl-shaped) point; **mirror** distinct, consisting of a rectangular to transversely elliptic or somewhat horseshoe-shaped figure (less frequently H-shaped or reduced to a few isolated speckles), usually centrally placed without connection to the base of the lip, shining (less frequently dull) (greyish) violet to bluish grey (rarely reddish brown), occasionally with a pale border. **Column** acute to obtuse, not tapering towards the base (in side view); stigmatic cavity at least as wide as long and approximately twice as wide as the anther, with dark (rarely pale), lateral, eye-like knobs at base. FIG. 96.

REFERENCES: Baumann & Künkele (1988), Berger (2003), Bournérias (1998), Buttler (1986), Caporali et al. (2001), O. Danesch & E. Danesch (1971, 1971a), Davies et al. (1983), Del Prete & Tosi (1988), Delforge (1989, 1994, 2001, 2005), Ehrendorfer (1980), Gölz & Reinhard (1975), Grünanger et al. (1998), Hahn & Salkowski (2005), Paulus & Gack (1986, 1990c, 1999), Rossi (2002), Salkowski (2000), Souche (2004), Sundermann (1980), Walravens (1995).

DUBIOUS RECORDS

Ophrys lutea subsp. *aspea*

Ophrys lutea Cav. subsp. *aspea* (J. Devillers-Terschuren & P. Devillers) N. Faurholdt

Syn. *O. lutea* Cav. subsp. *murbeckii* sensu H. Baumann & Künkele [non (H. Fleischm.) Soó].

Compared to the other subspecies of *O. lutea*, subsp. *aspea* (FIG. 97) is characterised by the fact that the lip of its relatively large flower is strongly vaulted and has a distinct knee-like bend close to the base. Furthermore, the lip exhibits a marked contrast between yellow and blackish brown areas together with a normally entire mirror. At the same time, its longitudinal furrow at the base is obscure.

This bee orchid has been reported from Sicily. We believe, however, that bee orchids from this island that have a superficial similarity to *O. lutea* subsp. *aspea* are hybrids between *O. fusca* and *O. lutea* s.l.

In the northern, Mediterranean part of North Africa (particularly Tunisia and Algeria), pronounced confusion surrounds *O. fusca* and *O. lutea*, partly because of frequent hybridisation. However, *O. lutea* subsp. *aspea* is readily recognised from the features mentioned above. [Not confirmed from Europe; Tun].

Ophrys reinholdii subsp. *straussii*

Ophrys reinholdii H. Fleischm. subsp. *straussii* (H. Fleischm. & Bornm.) E. Nelson

Syn. *O. straussii* H. Fleischm. & Bornm.

FIG. 97

FIG. 97. *Ophrys lutea* subsp. *aspea* (Tunisia, Djebel Zaghouan, 15th March 1998). Photo by H. Æ. Pedersen.

FIG. 98. *Ophrys reinholdii* subsp. *straussii* (Turkey, Anatolia, Antalya, Ibrâdi, 18th May 1995). Photo by
C. A. J. Kreutz.

Subsp. straussii (FIG. 98) differs from subsp. *reinholdii* in its smaller flowers with recurved lip margins and in the small mirror that is usually without connection to the lateral margins of the lip. In addition to the morphological characteristics, it flowers substantially later, with a peak in May.

This subspecies has been reported (but in all probability erroneously) from Chios and Rhodes. We are convinced that the individuals in question belong to subsp. *reinholdii* which occurs in an array of forms on those islands.

The distribution extends from the province of Antalya in southern Anatolia through to Iran and Iraq in the east. Additionally, this subspecies was recently found at one locality in Cyprus. [Not confirmed from Europe; Ana, Cyp].

Ophrys fuciflora subsp. *bornmuelleri*

Ophrys fuciflora (F. W. Schmidt) Moench subsp. *bornmuelleri* (M. Schulze) B. & E. Willing
Syn. *O. bornmuelleri* M. Schulze s.s.; *O. holoserica* (Burm.f.) Greuter subsp. *bornmuelleri* (M. Schulze) H. Sund.

Subsp. bornmuelleri (FIG. 99) differs from other subspecies of *O. fuciflora* in its nearly horizontally orientated, trapeziform lip, its usually upcurved (and often more or less yellow) lip margins and its very small (c. 1.5 × 2.2 mm long), knob-like petals. The sepals are green or whitish, less frequently rose-coloured. Flowering takes place later than in subsp. *grandiflora*, from late March (in the Levant) to late May (in southeastern Anatolia).

This subspecies has been reported from mountains on central Rhodes, but we are convinced that the plants in question belong to the highly variable *O.* ×*vicina* nm. "*heterochila*" which occurs abundantly in that area. Some of these small-flowered plants superficially resemble *O. fuciflora* subsp. *bornmuelleri*.

The distribution encompasses the Levant, southeastern Anatolia and Cyprus. [Not confirmed from Europe; Ana, Cyp].

Ophrys fuciflora subsp. *grandiflora*

Ophrys fuciflora (F. W. Schmidt) Moench subsp. *grandiflora* (H. Fleischm. & Soó) N. Faurholdt
Syn. *O. levantina* Gölz & H. R. Reinhard.

Ophrys fuciflora subsp. *grandiflora* (FIG. 100) resembles subsp. *bornmuelleri*, but differs in its square to cordate lip, which is pendent to slightly reflexed; in

FIG. 99. *Ophrys fuciflora* subsp. *bornmuelleri* (Cyprus, southern part, Pegeia, 18th March 2001). Photo by N. Faurholdt. FIG. 100. *Ophrys fuciflora* subsp. *grandiflora* (Cyprus, southern part, Pegeia, 18th March 2001). Photo by N. Faurholdt.

its small, often isolated mirror and in its more compact habit. Additionally, it flowers a couple of weeks earlier.

This bee orchid has been reported from Rhodes, but we are convinced that the plants in question belong to O. ×*vicina* nm. "*heterochila*".

The distribution encompasses Cyprus and southern and southeastern Anatolia, whereas *O. fuciflora* subsp. *grandiflora* has not been found with certainty in the Middle East. [Not confirmed from Europe; Ana, Cyp].

REFERENCES: Ashley et al. (1982), Baumann (1975), Buttler (1986), Devillers & Devillers-Terschuren (2000), Hervouet (1984), Hervouet & Hervouet (1998), Peter (1989), Saliaris (2002), Taylor (2005).

6. Bee orchids and conservation

Most bee orchids require fairly specific habitats and therefore belong to those groups of plants that have generally declined with the impoverishment of natural environments during the last century. Whether man should attempt to change such a trend, is always open to discussion, but there seems to be increasing political and public consensus that it is both ethically, scientifically and socially desirable to preserve biological diversity to the widest possible extent. Consequently, it is also relevant to scrutinise the current situation of the bee orchids in a conservation context.

Plant collecting poses a threat to a few particularly attractive bee orchids, for example *O. fuciflora* subsp. *lacaitae*, but, in general, damage or deterioration of habitats plays a much more important role. For instance, reclamation and construction work pose serious threats to several populations of entities such as *O. insectifera* subsp. *aymoninii*, *O. omegaifera* var. *basilissa*, *O. fuciflora* subsp. *chestermanii* and *O. fuciflora* subsp. *lacaitae*. Building in connection with residential tourism continues to expand,

Fig. 101. Expanding coastal urbanisation in southernmost Spain (Andalusia, Benalmodena Costa, 16th April 2001). Building in connection with residential tourism often destroys *Ophrys* habitats of great value. Photo by H. Æ. Pedersen.

especially in littoral regions of the Mediterranean (Fig. 101), and often completely destroys *Ophrys* habitats of great value. This poses a serious threat to, for example, *O. argolica* subsp. *aegaea*, *O. argolica* subsp. *lesbis*, *O. sphegodes* subsp. *sipontensis* and *O. fuciflora* subsp. *andria*. Smaller islands are particularly vulnerable in this regard, because the coastal region constitutes a high proportion of their total area. Many bee orchids are also sensitive to even comparatively modest alterations of their habitat. For instance, intensified forestry poses a local threat to *O. insectifera* subsp. *insectifera* and *O. fuciflora* subsp. *parvimaculata*, whereas excessive grazing (usually by goats) endangers numerous populations of entities such as *O. omegaifera* var. *basilissa*, *O. argolica* subsp. *aegaea*, *O. argolica* subsp. *lesbis* and *O. fuciflora* subsp. *andria*.

Because of the above problems (among others), a number of bee orchids are now threatened in certain regions, and some of the commoner species are strongly decreasing. Active conservation and monitoring measures are therefore urgently needed. Numerous other plant species are in the same situation as the bee orchids, including representatives of a diverse selection of other orchid genera – and since the threats to bee orchids and the threats to other European orchids are largely identical, it is natural that superior initiatives in the field of species-orientated nature conservation often focus on the orchids as a whole.

Internationally, two initiating, stimulating, and coordinating organisations of relevance to orchid conservation in the European region exist. One is the Secretariat for the Conservation of European Orchids (SCEO), established in 1996 by delegates from a number of European orchid study groups and societies. The other, more influential, one is the Orchid Specialist Group (OSG) under the aegis of the International Union for Conservation of Nature and Natural Resources/Species Survival Commision (IUCN/SSC). To provide a firm funding base for the work of OSG, an independent non-profit organisation, Orchid Conservation International, was recently established and became a registered charity under UK law in 2005.

CONSERVATION STATUS AND PRIORITIES

The surveys of threatened species known as red data lists are very important when putting different conservation measures in an appropriate order of priority. It is obvious that the quality of such lists utterly depends on the quality of the systematic classification of the organisms concerned, and a continuous, thorough research in systematics is therefore crucial for maintaining and improving our ability to prioritise conservation measures.

No bee orchids are included on IUCN's latest red data list based on the global conservation status of individual species (http://www.redlist.org). However, a considerable number of other red data lists exist that merely cover a single country or an (inter)national region and highlight species which are considered threatened within that particular area. It is beyond the scope of this book to include a complete survey of which bee orchids are assigned which status categories in which countries and regions – but the list would be comprehensive.

As mentioned above, the systematic classification of plants is of fundamental importance for the composition of red data lists. Given a particular classification, however, it is other parameters that govern which species are entered on the list and which status categories they are assigned. The most important factors are the past and present distributions of individual species, their current population sizes, their habitat requirements, their reproductive biology and population genetics as well as the factors threatening them. Among these parameters, the current distribution of species is in many respects the simplest to assess, and not surprisingly, this is where our knowledge is best. Most remarkably, mapping of bee orchid distributions have been conducted since 1976 in many regions of the Mediterranean as part of a special orchid mapping project under the international Organisation for the Phyto-Taxonomic Investigation of the Mediterranean Area (OPTIMA). The additional parameters are, in general, less thoroughly studied, and the papers have been published in a much more scattered way. However, existing knowledge about the various parameters is still more frequently being surveyed in "conservation papers" dealing with individual species. In the case of bee orchids, such reviews have been published, for instance, in the national Greek red data book (concerning *O. argolica* subsp. *argolica*, *O. scolopax* subsp. *rhodia* and *O. sphegodes* subsp. *helenae*), in the national British red data book (concerning *O. fuciflora* and *O. sphegodes*) and in an article on *O. fuciflora* subsp. *chestermanii*.

PROTECTION OF SPECIES OR HABITATS?

Bee orchids, like other plants, can be protected in two fundamentally different ways: either through specific protection of the living individuals of the species itself or through protection and management of its habitats.

Protection of the living individuals of a species most directly takes place through legislation, and legal protection of one or more *Ophrys* species already exists in several European countries (and national regions). Legal protection, in this context, usually means that individuals of the species in question may be neither picked nor dug up; on the other hand, it does not prevent damage to the habitats of the species. Historically, legal protection of species in Europe has been an entirely national matter. Today, however, the EC Habitats Directive obliges the federal partners of the EU to implement national legislation to protect species considered particularly threatened in the EU as a whole.

More indirectly, the species protection can take place through control and regulation of international trade in threatened plants. For this purpose, more than a hundred countries world-wide have ratified the Convention on International Trade in Endangered Species (CITES), also known as the "Washington Convention". The species covered by CITES are (depending on conservation status and commercial interest) included in Appendix I, II or III. All bee orchids are included in Appendix II. CITES operates on the definition that all transport of "appendix species" across frontiers can be categorised as international trade. It should be noted that the EU is usually regarded as one country in this context. In principle, material for scientific purposes is exempted (but, unfortunately, the practical precepts often appear unworkable, for which reason important research in threatened species may be unduly obstructed).

Legal protection of species and regulation of the international trade in bee orchids may benefit the very few bee orchids to which picking or digging up poses a real threat. Furthermore, these measures may have a more general, indirect effect – provided that they are accompanied by public relations campaigns. There is little doubt, however, that combined protection and management of natural bee orchid habitats is by far the best fitting key to the future existence of a rich *Ophrys* flora. Only through protection and well thought out management of the habitats can the really serious threats against bee orchids be counteracted. In addition, this strategy simultaneously improves the environmental conditions for many other species of plants and animals that are specially adapted to the same habitats.

Protection of individual localities traditionally takes place through legal nature conservancy, and today this is still the best way to provide reasonably certain long-term protection to natural habitats. However, the legal processing of nature conservancy cases is often prolonged and can be very expensive. Since, at the same time, there is a great and urgent need for protection and management of a high number of localities (including many that are too small for legal protection anyway), there is a widespread and increasing trend among public authorities in several European countries to enter into informal, inexpensive management contracts with the owners of localities worthy of preservation. Such contracts make it possible to protect and manage a high number of localities with short notice, but, being vulnerable to transfer of property, they provide only short-term protection.

Most of the typical *Ophrys* habitats are maintained by moderate grazing or forestry, mowing or recurrent burning; without this intervention the vegetation would gradually become dominated by more robust and competitive species. On the other hand, bee orchids are also vulnerable to excessive agricultural use of their habitats and will do best with a type of management that imitates the traditional ("old-fashioned") agriculture or forestry of the locality concerned. When nature management has to be arranged for the special benefit of particular species, it is obviously a great advantage to have a thorough knowledge of the ecology and biology of these species. Consequently, there is every reason to continue studying the natural history of bee orchids.

Historically, protection of habitats in Europe has been an entirely national matter, as with protection of species. Today, however, there are several international statutory instruments. For example, the EC Habitats Directive obligates the federal partners of the EU to carry through national protection and management of localities that represent habitat types (including some important *Ophrys* habitats) considered particularly endangered in the EU as a whole.

GENE BANKS, ARTIFICIAL PROPAGATION AND RE-INTRODUCTION

In recent years, artificial propagation and re-introduction of threatened orchids into the wild have gradually come to play a part in supplementing the conventional methods of protecting species and habitats described above. Up to now, only a few bee orchids have been included in such initiatives, but a number of *Ophrys* species are included in germination experiments under The Sainsbury Orchid Conservation Project at the Royal Botanic Gardens, Kew. Under the same project, re-introduction experiments have been performed with *O. apifera* – a species that botanists have also attempted to re-introduce to Schleswig-Holstein. It is probably just a matter of time before many more bee orchids are included in similar projects.

Propagation projects that aim at re-introduction to the wild are preferably based on growing plants from seeds. In contrast to vegetative propagation by meristems (cloning), propagation by seed involves the possibility of a high genetic variation among the raised seedlings. Seeds are usually collected in carefully chosen natural populations. The capsules are harvested shortly before ripeness and are dried in test tubes. The seeds can be sown immediately, or can be kept for a few years in a refrigerator. There is evidence that orchid seeds may keep their germination capacity for a long time if they are kept at low humidity in a freezer (-20 °C) or, perhaps even better, in liquid nitrogen (-196 °C); under these conditions they can probably stay viable for several decades, so it should be possible to establish long-term gene banks consisting of frozen seeds.

Artificial germination of orchid seeds is performed on nutritional media of a jelly-like consistency, either symbiotically (i.e. with participation of a mycorrhizal fungus) or asymbiotically. The majority of European orchids germinate most successfully if the symbiotic method is applied. In the case of bee orchids, however, fine results have also been achieved by the asymbiotic method, which is very similar to the symbiotic method, but takes place under completely sterile conditions and without participation of a fungus in the early stages.

The introduction of artificially raised seedlings into the wild is usually attempted when the plants are one to two years old. *Ophrys* is preferably planted out in the autumn, just as the plants start to shoot after the summer dormancy.

It should be emphasised that application of the new, progressive methods is controversial. Deliberately manipulating the plants, one constantly balances on the verge of "flora falsification" (i.e. deliberate "improvement" of natural habitats with foreign plants). It is, however, open to debate at which exact point one goes beyond the bounds of ethical propriety. Possible scenaria to consider from an ethical point of view include: (1) planting out seedlings artificially raised from seeds collected at the same site; (2) re-introducing the species to another site where a previous occurrence has become extinct and (3) introducing a species to potentially favourable areas (localities, regions, countries) where it has never occurred before. The IUCN/SSC Reintroduction Group has compiled a series of guidelines for re-introduction that can be reached through the homepage of IUCN, but today there are still great discrepancies between the political and ethical principles that different European countries follow. Personally, we feel that re-introduction programmes should in general be restricted to cases concerned with species that are threatened on the global scale. However, reinforcement of populations on a strictly local scale (i.e. with no geographic manipulation of genetic material) may occasionally be justified (as a supplement to habitat management) to preserve particularly isolated and threatened gene pools of species that are not necessarily threatened in their main distribution areas.

2

MONITORING OF POPULATION DEVELOPMENT

There is a constantly increasing realisation that systematic monitoring of the development in populations of threatened species must be incorporated as a necessary component in conservation projects. Monitoring can reveal if a previously healthy and thriving population encounters a problem so serious that a more focused habitat management (and perhaps even direct assistance through artificial propagation) is called for. However, monitoring plays an equally important role as a feedback tool which reveals the (positive or negative) reaction of populations from habitats where management measures have recently been implemented or changed.

Monitoring of population development (FIG. 102) can be performed in numerous ways — one can: monitor the entire population or lay out permanent plots or transects; count all individuals or only the flowering ones; count individual shoots or vegetatively coherent clones; do the counting at different seasons and with very different time intervals; follow the fate of individual plants from one year to another, or only deal with the total number of individuals each year. It generally applies that the more time-consuming and thorough methods you use, the more detailed and informative results you will get.

The most thorough monitoring projects on *Ophrys* populations to date

FIG. 102. Monitored population of *Ophrys insectifera* in Wales (Anglesey, Cors Bodeilio, 12th June 1999); each specimen is marked with a stick. Monitoring of population development is an important feedback tool that should be incorporated in all conservation and management projects aimed at particular species. Photo by H. Æ. Pedersen.

were conducted on *O. sphegodes* in England by M. J. Hutchings and S. Waite. Individual plants were followed from the first time they appeared above the ground. The very detailed results made it possible to develop mathematical, so-called matrix models. These can be utilised for simulating the long-term population development under each of the management forms that have been applied to the locality during the period of monitoring. Additionally, the models can be used for selecting those life stages of the bee orchid to which altered management measures can be most successfully directed in order to effect a positive population development (in this context, a close coherence with the general life-history strategy of the species must be expected). Development of similar matrix models for populations of additional *Ophrys* species is highly desirable, because it must be assumed that their information concerning efficiency of different management measures can to a certain extent be applied to other populations of the same species.

REFERENCES: Baumann & Künkele (1995, 1995a, 1995b), Baumann et al. (1995), Bjørndalen (2006), Del Prete (1999), Delforge (1996), Dorland & Willems (2002, 2006), Fay & Krauss (2003), Frosch (1983), Hágsater (1996), Horsfall & Wigginton (1999), Hutchings (1987, 1987a, 1989), Kapteyn den Boumeester (1997), Kell et al. (2003), Koopowitz (2001), Lavarack & Dixon (2003), Lucke (1971), Malmgren (2006), Muir (1989), Neto & Custódio (2005), Ramsay & Dixon (2003), Ramsay et al. (1994), Reinecke (1995), Richard & Evans (2006), Roberts (2003), Ronse (1989), Salanon & Kulesza (1998), Stewart (1987, 1992), Stone & Taylor (1999), Waite (1989), Waite & Hutchings (1991).

7. *Ophrys* cultivation

Successful cultivation of *Ophrys* requires knowledge and understanding of the growth cycle of these plants, and how to treat them at different stages. Starting with the summer-dormant tuber (FIG. 103), growth begins in autumn, typically in late September or October, exceptionally earlier or later, depending on the outside temperatures and humidity. A few nights of sub 10 °C temperatures often provokes the plants into growth. Growth seen at the soil surface proceeds as a pointed green bud or "nose" which soon expands into a leafy rosette; at the same time the roots start spreading from the stem immediately above the tuber. By November or December a specialised "dropper" root appears, thicker than the others and descending vertically. Its tip gradually enlarges to form a new tuber, which is mature by flowering time. The flower spike emerges from the centre of the rosette, and it can be spotted in its initial stages as early as December or January. During or soon after flowering time the leaves wither, yellow and die back; a few species such as *O. insectifera, O. apifera,* and *O. fuciflora,* in the north of their range, may delay this die-back if the ground remains moist and they are not heat-stressed.

FIG. 103. *Ophrys tenthredinifera*, dormant plant showing newly matured tuber and withering remains of old plant. Photo by R. L. Manuel.

Fig. 104 Flowering plants of *Ophrys speculum* subsp. *speculum* in pot. Photo by R. L. Manuel.

Until recently the only sources of *Ophrys* plants for growers were of plants originally taken from the wild, although the suppliers often tried to disguise this fact. Thankfully, such plants are now in diminishing supply, largely due to the development of methods of growing these plants from seed. The Journal of the Hardy Orchid Society (http://www.hardyorchidsociety.org.uk) attempts to include only advertisements from nurseries that offer legitimate, seed-grown plants.

Supplementary information on many aspects of culture and seed growing of *Ophrys* can be found in a forthcoming book tentatively entitled *Cultivation of Temperate Orchids* (R. L. Manuel, in prep.).

GROWING ENVIRONMENT

The majority of growers will house their *Ophrys* plants in either a greenhouse or a cold frame. *Ophrys* have a surprising resistance to frost (down to -10 °C) but must be kept fairly dry at low temperatures. Within the confines of a greenhouse or cold frame the air can become saturated with moisture, which will encourage the growth of *Botrytis*. As *Botrytis* can rapidly kill the plants, plenty of ventilation is necessary, and unless the weather is extremely cold (sub-zero) a free flow of outside air, aided by an electric fan, is essential.

Ophrys can also be grown indoors in a suitably cool (unheated) room, a cellar being ideal. An artificial light source is necessary and the wide spectrum fluorescent tubes used by aquarists are ideal. One of these tubes together with an ordinary warm white tube, mounted side-by-side 30–40 cm above the plants and beneath a good reflector to direct the light downward, is adequate. They should be set to a day length of 12 hours during September and March, 10.5 hours in October and February, and 9 hours for November, December and January, using a timer. Such a set-up will provide sufficient light for a block of pots about 40 cm wide and as long as the tubes. Ventilation and air movement are as important in this environment as in the greenhouse.

POT CULTURE

Ophrys are normally grown in pots (FIG. 104). The ideal is to use small clay pots (a 3–4 inch pot for one mature plant is about right) plunged into damp sand, but this is not economical with space (FIG. 105). Plastic pots are quite satisfactory provided the grower understands that plastic pots retain water much longer than clays, and that free standing pots are liable to heat up in sunlight.

The potting composts used must be very free draining. Their exact constitution is relatively unimportant but should consist of an organic component plus a high proportion of inorganic matter for drainage, texture and pH control. For the latter, sharp grits and manufactured inorganic materials such as perlite or the various baked clay granules used in horticulture are useful, as is pumice gravel. The organic fraction of the compost provides the main food source. It should consist partly of loam, either fibrous or with a good crumb structure (loam which readily reduces to a powder should be avoided) and partly other organic materials, such as good quality hard leaf mould (beech or oak, not the wet slimy product of other European trees), composted bark as chips or fibre, or pine duff from a not too acidic site. Probably all *Ophrys* prefer soil with an alkaline reaction (pH > 7.0) and to ensure this a modest amount of calcareous shell grit, or limestone grit, should be incorporated (not horticultural lime, which may burn the roots). Otherwise, as the compost rots down during the course of the plant's life cycle, it would naturally acidify, which is detrimental to the development of the new tuber. The structure of the compost also

FIG. 105. *Ophrys* (and other genera) growing in clay pots plunged into sand bed in greenhouse, December. Photo by R. L. Manuel.

deteriorates during the annual cycle, so plants should be potted into fresh compost each summer.

The following mixture is used with success by many British growers but Continental growers often use more inorganic mixtures, with as much as 70–80% inorganic materials. This has the advantage of even better drainage but dries out readily, so watering must be very frequent and close attention given to the moisture content of the pot, or the plants may suffer, especially in unexpected warm or windy spells. A typical *Ophrys* compost mixture consists of:

- 25% good loam, preferably from a calcareous source, as described above.

- 25% beech (or oak) well rotted leaf mould, which should have a light fluffy texture.

- 50% (or more) inorganics. 40% (or more) medium grit (3–6 mm), 10% calcareous grit ('poultry grit', as sold by agricultural merchants, is perfect).

In addition, up to 25% perlite can be added to the mix; this opens the compost even more and prevents the mixture becoming too heavy and compacted.

'Good drainage' means that when water is poured into the compost from above, most of it runs out through the drain holes of the pot, leaving the contents no more than moist and not much heavier than before. Overwatering, often blamed for dead plants, should be impossible if suitable compost is used. Orchids, wherever they grow, love to have well-aerated water running over their roots, but sitting for long periods in stagnant water (or of course, wet compost) with a low oxygen content, will soon finish them off, especially in cold weather. Remember, these are winter-growing plants.

Problems of winter growing usually arise not just from cold itself, but from a combination of cold (when the plants will only be growing very slowly, and thus using little water) and over-wet compost. During December and January in particular, watering should be kept to a minimum. It is also important to anticipate cold weather and reduce watering well before it arrives. As stated above, good ventilation is essential, especially during cold windless weather with dank humid air, which is when *Botrytis* or related fungal rots are most likely to strike. As soon as signs of spring appear and the weather warms up, the plants will increase their growth rate and will need more water. Despite their resistance to moderate frost, it is much easier and safer to grow the plants, in the greenhouse, frost-free. This is easily achieved using a small electric fan-heater, and the running costs are low. Many of the potential problems of winter growing are reduced, but unfortunately not entirely eliminated, by this means.

When the plants have flowered and the leaves start to die back watering should cease and the pots allowed to dry out completely. Once dry it is a good idea to empty the pots to inspect and admire the tubers. These can be stored in a cool dry place during the summer – a summer baking would cause them to shrivel. They are probably safer buried in the dry old compost until signs of new shoots appear as autumn approaches. They should then be repotted in fresh compost. Because the roots arise above the tuber, the latter should be buried at least its own length below the surface, two or three times this for small seedlings. Some growers prefer to leave their newly potted plants out in the open, to be watered by rain and to experience cooler nights; others prefer to withhold water until new shoots appear above the surface of the compost, indicating that root growth has started. Both methods have certain advantages; it is largely a matter of choice and convenience.

PESTS AND DISEASES

Two potential threats to the well-being of *Ophrys* plants stand out above all others. Many of the usual greenhouse pests are inactive during the winter, and so need not be considered here.

Botrytis has already been mentioned. Unfortunately, once the all too familiar greyish fuzz of fruiting bodies appears on leaves or other parts (which soon turn black) the infection is well established and can only be eliminated by cutting away the infected part. If the infection reaches the stem, which typically happens at compost level, the plant is, effectively, killed, as the stem is severed and the plant will not initiate new growth. The use of systemic fungicides might be beneficial, but may also damage or kill any beneficial mycorrhiza involved with the orchid.

The other enemy is the greenfly (aphid). These well known small insects operate at reduced activity during the cold weather of winter, but the slightest hint of a warm spell will cause them to increase their numbers almost overnight. They will lurk in small but increasing numbers under the rosette leaves, or occasionally nestle right down in the centre of the rosette, where the newest, lushest plant tissue is emerging. The only preventive is eternal vigilance – regularly inspect the plants, turning leaves up to look underneath, and be aware that any slight yellowish or pale spotting on the leaves may signal an aphid colony on the underside. Once found the aphids should be

FIG. 106. Various bee orchids growing on a grassy bank in author's garden. **A.** *Ophrys omegaifera* subsp. *dyris*, the same plants as in B, in flower; **B.** *Ophrys omegaifera* subsp. *dyris*, four plants derived naturally by vegetative division from a single original; **C.** *Ophrys reinholdii* and *O. kotschyi* subsp. *cretica* in flower on the same grassy bank. Photos by R. L. Manuel.

despatched instantly with a suitable insecticide; pyrethroids are very effective and, although not systemic, do tend to discourage further colonisation of a sprayed plant. Once spring arrives aphids can be found on any part of the plant, especially on fresh flower buds. This usually causes a distortion of the flowers when they open, so even greater diligence is required.

OPHRYS IN THE GARDEN

Ophrys are not ideal garden plants – they are too small for normal flower beds, and being summer dormant, are easily overgrown by more vigorous summer growers. But with great care and sensible preparation, they can be grown in suitable sites such as on a steep grassy bank with short turf (which will have to be cut by hand with care!). The perfect soil would be calcareous and well drained, but this writer has a bank with heavy clay soil where many *Ophrys* thrive, as the slope is sufficient to shed even continuous heavy rain. Plants of various species have been grown here for up to six years so far, and while a few have been lost, others have multiplied in this situation (FIG. 106). Ideally the plants should be exposed to as much sun as possible during their winter growth period. They should be planted out as dormant tubers towards the end of the summer, by cutting out a plug of turf and then replacing it over the tuber. Frost down to about 10 °C should not be a problem; probably the greatest danger to the plants is that molluscs or rodents might eat them. An alternative growing site is an alpine trough, but this may need to be covered against rain if the drainage is less than perfect. Also remember that during the summer the tubers like to be mostly dry, so if other plants are grown in the trough it would be better to remove and store the *Ophrys* tubers.

VEGETATIVE PROPAGATION

Only one *Ophrys* species, *O. bombyliflora*, habitually makes more than one tuber per year. Others may naturally make an extra, small tuber if mature and well grown, but this is exceptional. It is possible to divide plants in the hope that a secondary tuber can be induced, but this is risky and may result in the loss of the plant.

The (usually) short root-tuber between the old stem and the nearly mature new tuber is severed; this should be done around flowering time. The process requires that the plant is partly bare-rooted, to expose the relevant parts (FIG.107). Treat any cut surfaces with flowers of sulphur before repotting. After surgery the plants should be kept in growth for as

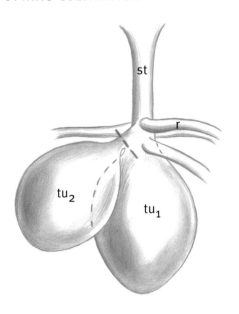

FIG. 107. *Ophrys* underground organs, showing method of vegetative division. **tu$_1$**: old tuber; **tu$_2$**: young tuber; **r**: root; **st**: stem. The broken line shows where to cut. Drawing by R. L. Manuel (redrawn by Linda Gurr).

long as possible in a cool shady place, under the greenhouse bench, for example, and with luck this may result in one or more tiny new tubers being formed on the old stem. The separated new tuber should be kept cool and slowly allowed to become dry.

GROWING FROM SEED

This is a huge subject and there is only space here for a brief review of what has been achieved to date. *Ophrys* are relatively easy to grow from seed using either the symbiotic method (with a suitable mycorrhizal fungus) or asymbiotically, using a nutrient medium to encourage germination and to feed the developing plantlets. Various techniques of sterilising, sowing, replating, and growing on the seedlings can be found in the cited literature.

The best seed is produced by hand pollination between two different parents of the required species. Hand pollination in this genus is easy as the pollinaria are large and easily handled by using fine forceps to grasp the bursicle (not the viscidium) and press the pollen onto the stigma. Self-pollination usually produces poor seed with a low embryo count, although *O. apifera* is habitually self-pollinating and usually produces large quantities of good seed. Seed from the current year is often reluctant to germinate, but germinates freely in subsequent years. Once cleaned and thoroughly dried, store the seed in a refrigerator at about 4 °C. Good seed stored in this way retains its viability for at least six years, and some seed of ten years old has been germinated and grown, though at a lower viability level.

The asymbiotic germination method, which is similar to that used for

Fɪɢ. 108. Typical developmental sequence of *Ophrys* germinated in symbiotic culture. Sequence from seed to seedling near end of first season's growth. Proportions may vary considerably. **A–B.** In June/July the embryo imbibes water and swells, while the seed coat gradually decomposes; **C.** In August the protocorm forms and polarises, root hairs appear, and fungal pelotons can be seen as dark internal bodies at the basal end; **D.** In August–September the first leaf initial can be observed; **E–F.** In September–November the first roots emerge; **G.** From December onwards the first tuber develops. Drawing by R. L. Manuel (re-drawn by Linda Gurr).

tropical orchids, is relatively easy but has certain disadvantages; for instance, weaning the seedlings can result in large losses. The abrupt change of environment, including the stark change in nutrition, seems to be a difficult hurdle to overcome for many plants. Generally, these seedlings should be grown *in vitro* for a longer period than symbiotic plants, until the tubers are well formed. This gives them a better chance of survival when de-flasked. Asymbiotic growth media are relatively complex and their ingredients may be difficult for an amateur to obtain or to make up. Recommended media are BM1, and SM (Svante Malmgren medium: but omitting kinetin and substituting potato or banana for the pineapple juice in post-germination growth media – pineapple is good for germination but seems to have an inhibitory effect on tuber maturation). Sigma Phytamax™, at half strength plus 25 g pulped banana per litre, is excellent for replating and growing on.

Symbiotic germination and growing on, when a suitable fungus can be found, can work extremely well. The seedlings are easy to wean as the fungus continues to provide nutrition, resulting in less traumatic stress during the change of environment. The growth medium used for this (oats medium) is relatively simple and the ingredients easy to obtain. The major problem is obtaining a suitable mycorrhiza. None appears to be available commercially, so a fungus must be extracted from existing plants. Not all the fungal isolates from an orchid's roots will germinate or grow seedlings, and each one has to be tested rigorously before it can be used.

Ophrys seedlings follow the same developmental sequence as their close relatives *Anacamptis* and *Serapias* (*Orchis* is rather less consistent in development). Development from seed to young plant with its first tuber is shown in Fig. 108. This has proved to be very consistent throughout the genus for seedlings grown symbiotically. Asymbiotic seedlings tend to grow roots that are thicker and longer, especially the dropper root, which often pushes the whole plant clear of the growth medium in the flask, and abnormalities occur fairly frequently with this method.

Symbiotic seedlings should be de-flasked into pots once the roots are at least 1 cm long, preferably more. There is usually a dropper root visible by this stage, which ideally should be timed to occur in January or February. Open the flasks for about one week before de-flasking, admitting fresh air to allow the leaves to harden; then plantlets can be potted into normal compost (as described above) in community pots. The seedlings should be grown on in shaded and humid (never arid) conditions until the leaves wither, signalling dormancy in the new tuber. They can then be treated as adults, although they may be later to appear the following autumn than their parents. Asymbiotic seedlings should be kept in the flask longer, until their tubers are well formed but while the leaves are still green and not yet showing signs of going over, typically around March or even April.

Post-flask development differs between cultivated seedlings and those which grow in natural conditions. In culture environmental conditions are stable, nutrition is constant and consistent, and pests and competition from other plants are eliminated, whereas none of these necessarily applies in natural conditions. Cultivated seedlings typically produce a leaf rosette of (1–)2–4 (rarely more) leaves in their first year, and tubers can be as large as 1 cm long (larger in asymbiotic seedlings) (FIGS 109–110). In their second year the rosette can be almost as large as in flowering plants, and a few seedlings may flower (flowering is probably impossible at the end of the first year, as there is no old tuber to support the necessary growth surge). Many plants flower in their third and subsequent years, although a few species, such as *O. tenthredinifera*, seem to take longer to flower.

In wild populations of many species, observed over 20-plus years of orchid holidays (and in the garden), it is clear that wild seedlings usually only make one small, rather narrow, leaf in the first year. This is probably because they germinate later than cultivated plants, and life in the wild is harder! Second year plants can be almost as large as cultivated ones, though few flower at this age.

Richard L. Manuel

REFERENCES: Malmgren (2002, 2004), Manuel (1996–1997, 2005), Mitchell (1989), Ramussen (1995), Seaton & Ramsay (2005), Thompson (1977).

(left) FIG. 109. *Ophrys tenthredinifera*, root system of young plant around December. Photo by R. L. Manuel.
(right) FIG. 110. *Ophrys tenthredinifera*, seedlings near end of first year's growth. Photo by R. L. Manuel.

8. Recommended *Ophrys* excursions

Ophrys occurs throughout most of Europe, but in the search for plenty, or for particular species, the Mediterranean region offers outstanding opportunities (FIG. 111). Bee orchids can frequently be found growing in close proximity to popular hotels or along paths in those natural areas or archaeological sites that constitute the traditional destinations of coach tours arranged by tour companies. However, the chances of finding large populations (or rare species of bee orchids) can be greatly improved by systematically searching for them in particularly favourable places (see the section *Ophrys* habitats p. 34). If you have a particular aim, such as finding an especially rare and narrowly distributed bee orchid, considerable preparations are often required, as well as a fair share of luck.

The committed *Ophrys* enthusiast rents a car and often chooses to move from place to place along the route rather than to stay in the same hotel. In this way, a considerable number of *Ophrys* species can be seen during a trip of one or two weeks. Still, a more traditional visit to the Mediterranean may offer great experiences, too.

Among the destinations we have visited ourselves, a number of

particularly recommended areas are briefly described below, starting from the west, and working eastwards. It should be noted that the list includes destinations to which no organised tours are arranged, and others where the charter season does not start until the peak flowering of bee orchids has passed. Indeed, it is worthwhile remembering that the optimum period for *Ophrys* studies in most of the Mediterranean is generally from early March to early May, which means that the traveller will frequently have to book scheduled flights.

FIG. 111. Precious moments from *Ophrys* excursions in the Mediterranean – joy, excitement and wonder (*Opposite page* **A.** Greece, Rhodes, Trianda, 11th April 1993; *top left* **B.** Greece, Rhodes, Kamiros, 15th April 1996; *top right* **C.** Italy, Puglia, Monte Gargano, 12th April 1995; *left* **D.** Italy, Sardinia, Dorgali, 22nd April 1997). Photos by N. Faurholdt.

FIG. 112. Numerous fine *Ophrys* localities surround the city of Loulé in the Algarve. In this *Cistus*-dominated garrigue south of Paderne, several species and subspecies of bee orchids can be found, including *O. lutea* subsp. *galilaea* that has only been recorded very few times from Portugal. Photo by N. Faurholdt, 4th April 1999.

THE ALGARVE

The Algarve forms the southwesternmost part of the European continent. An early spring is responsible both for an early start to the tourist season and for the rather built-up coastal landscape. West of the city of Lagos, it is still possible to botanise on unspoiled coastal hills, but in the Algarve region you will have to go north, 5-10 km away from the coast. Here, the landscape appears as a mosaic of small neglected agricultural fields and unreclaimed *Ophrys*-rich garrigue (FIG. 112), but also incorporates expanding *Citrus* plantations. Some of the best places are areas north and west of Lagos, in garrigue and abandoned fields around Loule, on roadside banks and calcareous hills north of Boliqueime and the only "mountain" of the Algarve: Rocha de Pena.

If you want to completely escape the noise and dust of the coastal region, there are rich opportunities for suitable accommodation in several charming cities further north, amongst them Loule, Alte, Silves and Vila do Bispo.

The most frequently occurring bee orchids in the Algarve are *O. lutea* subsp. *lutea* and *O. speculum* subsp. *speculum*, closely followed by *O. bombyliflora*. *Ophrys lutea* often occurs abundantly and in its most beautiful and large-flowered form. Special western Mediterranean bee orchids to be found in the province of Algarve include *O. speculum* subsp. *lusitanica* and *O. omegaifera* subsp. *dyris*. The former grows here and there, whereas the latter is fairly rare and very local. Finally, two subspecies of *O. scolopax* are found in the area, the more frequent of them being the small-flowered subsp. *apiformis*, which is at its height in the first half of April.

REFERENCES: Tyteca (1997, 2000), Tyteca & Tyteca (1986).

ANDALUSIA

To most people, Andalusia is synonymous with the Costa del Sol, an inferno of noise and heavy traffic, completely lacking in charm. The infernal part, however, is confined to a narrow belt along the coast, and moving uphill behind this belt you will find fine and varied landscapes (FIG. 113) that are only locally disfigured by modern, large-scale olive plantations.

The bee orchids most likely to be found here are *O. lutea* and *O. speculum* subsp. *speculum*, both of which are very common throughout most of the region. In light-open pine woods between Coin and Alhaurin de la Torre you may find *O. omegaifera* subsp. *dyris*, the only representative of *O. omegaifera* in the western Mediterranean. Even more interesting, this area also holds some of the best European populations of *O. atlantica*. Outside North Africa (Algeria and Morocco) this species is only found in Andalusia but, unfortunately, it is frequently dug up by orchid collectors – especially in the neighbourhood of Coin.

At the centre of a most beautiful landscape in Serrania de Ronda, about 60 km northwest of Marbella, one finds the old, picturesque city of Ronda,

FIG. 113. Zahara is one of many inviting small cities in Andalusia and one of several suitable bases for excursions into the *Ophrys*-rich hilly landscapes of southernmost Spain. Photo by H. Æ. Pedersen, 18th April 2001.

253

which offers very good hotels at reasonable prices and constitutes a perfect base for excursions into one of the most orchid-rich parts of Andalusia. *Ophrys scolopax* subsp. *scolopax* and subsp. *apiformis* are fairly frequent, and exceptionally large-flowered forms of *O. tenthredinifera* can be found with a bit of luck. Very locally, *O. bombyliflora* grows gregariously on roadside slopes and moist meadows.

REFERENCES: Ackermann & Ackermann (1992), Claessens (1992), Hertel (1989), Lowe (1998), Molero Mesa et al. (1981), Riechelmann (2006).

MALLORCA

Most tourists in Mallorca are based in the cities of Palma or Alcudia, both of which can serve as fine points of departure if you want to study the bee orchids of this island. The mountain range along the northwestern coast (FIG. 114) and the warm, dry coastal plains by the bays of Alcudia and Palma offer the best opportunities. On the coastal plains, which are dominated by garrigue and light-open pine forest, spring comes early, and the flowering of bee orchids will quickly cease on the dry, sandy soil. Between Alcudia and Can Picafort, *O. tenthredinifera* and *O. speculum* subsp. *speculum* can be found in large numbers, and it is not uncommon to witness pollination of the latter. *Ophrys lutea* subsp. *lutea* and *O. sphegodes* subsp. *atrata*, both of which are rare in Mallorca, also occur in this area.

The most frequent bee orchids are *O. bombyliflora*, *O. fusca* subsp. *fusca* and *O. speculum* subsp. *speculum*. *Ophrys fusca* subsp. *fusca* is represented with both large-flowered and small-flowered forms. *Ophrys tenthredinifera* is another common species in Mallorca, and it is not unusual to encounter individuals with remarkably large, colourful flowers or with strongly reflexed sepals.

Castillo de Bellver, surrounded by a beautiful neglected park, is situated on the outskirts of Palma. Several orchid species occur spontaneously in the park; for instance, a large population of *O. apifera* can easily be observed. *Ophrys omegaifera* subsp. *dyris* and *O.* ×*brigittae* grow in scattered, warm coastal scrubs southeast of Palma (FIG. 12) – together with *O.* ×*flavicans* which in Mallorca often has a three-lobed lip. The latter hybrid complex can also be found along a number of roads, for example between Andraitx and Puigpunent west of Palma, and in the park at Castillo de Bellver.

REFERENCES: Beniston & Beniston (1999), Blatt & Hertel (1984), O. Danesch & E. Danesch (1964), Hoffmann (1983), Krämer & Krämer (1992, 1996), Kreutz (1989).

FIG. 114. In the mountain range along the northwestern coast of Mallorca you may be accompanied by feral pigs while searching for bee orchids, as here in the neighbourhood of Coma Freda. Photo by H. Æ. Pedersen, 4th May 1989.

SARDINIA

More than one visit is required to cover all of Sardinia, one of the largest islands in the Mediterranean. Our first journey to this island included the Ogliastra region in its eastern part (FIG. 111D) which includes the most prominent mountain range of Sardinia, the Gennargentu massif, the highest peak of which reaches 1834 m altitude. It also accommodates by far the major share of Sardinian *Ophrys* species. A charming hotel in the small mountain city of Lanusei is perfectly situated for excursions in this region.

Commonly occurring bee orchids include *O. tenthredinifera*, *O. lutea* subsp. *galilaea* and various forms of *O. fusca* subsp. *fusca*. Locally, *O. ×arachnitiformis* can be seen in great numbers. The main speciality of Sardinia is the imposing, endemic *O. fuciflora* subsp. *chestermanii* which Sardinian friends showed us near Baunei and at a site north of Tertenia. The largest populations, however, were those that we saw in shady scrubs of holm oak north of Grotta di San Giovanni. This place, which also holds the large-flowered hybrid *O. fuciflora* subsp. *chestermanii* × *tenthredinifera*, is situated east of Iglesias in southwestern Sardinia. The late-flowering *O. scolopax* subsp. *conradiae* is considered endemic to Sardinia and Corsica.

An exposed limestone plateau west of Ulassei is home to *O. apifera, O. speculum* subsp. *speculum* and *O. sphegodes* subsp. *atrata* (as well as many of the bee orchids mentioned above). Areas along the road between Urzulei and Dorgali hold several kinds of bee orchids. In addition to *O. ×arachnitiformis* (which is highly variable at this place), *O. fuciflora* subsp. *fuciflora* and *O. fusca* subsp. *iricolor* should also be mentioned. The former we also found in the southeastern part of Isola di S. Antioco by the southwestern corner of Sardinia. We recommend visitors to this part of Sardinia to stay at "Betty's Bed & Breakfast" in Villamassargia south of Domusnovas.

REFERENCES: Giotta & Piccitto (1990, 1991), Gölz & Reinhard (1990), Hennecke & Hennecke (2000), Klinger (1974), Scrugli (1990), Scrugli & Cogoni (1998).

SOUTHWESTERN TOSCANA

At the southwestern corner of Toscana, about 150 km northwest of Rome, there is a small peninsula consisting of the maquis-covered Monte Argentario (c. 640 m), an area well-known and frequently mentioned because of its rich orchid populations. During our visit, we stayed at a hotel in the charming little city of Porto Santo Stefano on the northern coast (FIG. 115).

There is hardly any grazing by sheep or goats on Monte Argentario, which is, as a result, quite overgrown by maquis (FIG. 9). Consequently, the

FIG. 115. The idyllic city of Porto Santo Stefano on the northern coast of Mone Argentario is the perfect point of departure for *Ophrys* excursions in southernmost Toscana. Photo by N. Faurholdt, 7th April 2001.

bee orchids should mainly be looked for along the many small roads as well as in pockets of garrigue and on olive terraces. The specialities of this peninsula are *O. argolica* subsp. *crabronifera* and *O. sphegodes* subsp. *litigiosa* var. *argentaria*. The former is predominantly found along the small road leading to the telegraph station on top of the hill. Var. *argentaria* is seen scattered (but in places abundantly, FIG. 18) along roads all over the peninsula. Other bee orchids to be found in the area include *O. lutea*, *O. bombyliflora* and *O. sphegodes* subsp. *sphegodes*, as well as large-flowered and small-flowered forms of *O. fusca* subsp. *fusca*.

On the mainland, a little north of Monte Argentario, lies the national park of Maremma with a number of fine sites where bee orchids grow in profusion. In the neighbourhood of Albarese, in particular, there are olive groves where *O. bertolonii*, *O. tenthredinifera*, *O. sphegodes* subsp. *passionis*, and *O. sphegodes* subsp. *atrata* flower in April.

Further north you should visit the area around Castagneto to look for *O. ×arachnitiformis* and *O. fuciflora* subsp. *fuciflora*, both of which we found on roadside slopes. Several of the bee orchids mentioned above likewise occur in this area.

REFERENCES: Del Prete & Tosi (1981), Del Prete et al. (1993).

SICILY

Most charter hotels in Sicily are situated on the east coast between Catania and Taormina and offer numerous possibilities for excursions to the *Ophrys*-rich areas surrounding Etna. Southeastern Sicily, particularly the part slightly west of the historically important sea port of Siracusa, is even better. Here you will find a limestone massif, Monte Iblei, that accommodates large quantities of bee orchids in a landscape varying from idyllic to dramatic. In the small mountain city of Bucheri (FIG. 116), Hotel Monte Laura offers simple, but charming accommodation right in the centre of the area.

On calcareous grassland and in garrigue around Bucheri, you should be able to find the Sicilian endemics *O. fuciflora* subsp. *biancae* and *O. lunulata*, which are fairly common in this area. In the wider surroundings (FIG. 13) you might also encounter the South Italian specialities *O. fuciflora* subsp. *oxyrrhynchos* and *O. fuciflora* subsp. *lacaitae*. The latter is very rare in Sicily, but can still be found, for example, at a site near Necropoli di Pantalica and in the lowland close to Cassabile, occasionally together with *O. fuciflora* subsp. *candica* and putative hybrids between the latter and subsp. *oxyrrhynchos*. *Ophrys bertolonii* frequently accompanies the above species.

The most common orchids in Sicily are *O. lutea* (subsp. *lutea* and subsp. *galilaea*), *O. fusca* subsp. *fusca*, *O. sphegodes* subsp. *atrata* and *O. tenthredinifera*. Locally, especially north of Etna, *O. sphegodes* subsp. *passionis* is also common.

Ophrys omegaifera subsp. *hayekii* is one of the rarest members of the Sicilian orchid flora. This bee orchid seems only to be extant in the central and southeastern parts of the island, for example around Lago San Rosalia north of Ragusa. By Ficuzza, south of Palermo, the near-endemic *O. fusca* subsp. *pallida* grows (sometimes in fairly large numbers) in light-open scrub and stunted forest grazed by domestic animals (FIG. 10).

REFERENCES: Ackermann & Ackermann (1988), Bartolo & Pulvirenti (1997a, 2005), Falci & Giardina (2000), Galesi (1995, 1999), Grasso & Grillo (1996), Hertel (1984), Kajan (1987), Künkele & Lorenz (1995).

FIG. 116. The picturesque city of Buccheri, west of Siracusa in southeastern Sicily, is literally surrounded by first-class *Ophrys* sites. Photo by H. Æ. Pedersen, 17th April 1999.

MONTE GARGANO

Monte Gargano (FIG. 111C) is located in northern Puglia and constitutes the peninsula that forms the spur of the Italian "boot". This promontory is one of the most species-rich areas in Europe as far as orchids are concerned. There are no charter tours to Monte Gargano, so one needs to fly to Napoli or Bari and proceed in a rented car. There is plenty of accommodation in the small but impressive cities along the picturesque coast, although many places do not open until the flowering season of *Ophrys* has passed.

Monte Gargano holds populations of a few bee orchids that are special for central and southern Italy. *O. fuciflora* is represented by, for instance, subsp. *apulica* and subsp. *parvimaculata*, which are almost restricted to Puglia. The former occurs scattered throughout Monte Gargano, whereas the latter

is mainly found between Lago di Varano and Lago di Lesina. In the early spring, *O. ×arachnitiformis* flowers in the coastal pine woods. *Ophrys bertolonii* is comparatively rare, whereas *O. ×flavicans* is fairly frequent. *Ophrys argolica* subsp. *biscutella*, though very variable, can be easily recognised and flowers in April. Together with *O. sphegodes* subsp. *passionis* it is relatively common. *Ophrys sphegodes* subsp. *sipontensis* is endemic to the southern part of the promontory. *Ophrys scolopax* subsp. *cornuta* has its westernmost occurrences (and the only ones in Italy) along the north coast of Monte Gargano. However, the *Ophrys* flora of Monte Gargano is not all specialities: more widespread bee orchids that are common here include, for example, *O. lutea* subsp. *lutea*, *O. lutea* subsp. *galilaea*, *O. bombyliflora*, *O. tenthredinifera* and *O. sphegodes* subsp. *atrata*. Strange to say, *O. fusca* subsp. *fusca* is only found close to the southern coast.

REFERENCES: Andrew (2002), Lorenz & Gembardt (1987), Rossini & Quitadamo (1996).

FIG. 117. The unreclaimed limestone area Le Murge south of the sea port of Bari is home to a number of special southern Italian bee orchids; they thrive, for instance, in open oak forests such as this wood between Alberobello and Noci. Photo by N. Faurholdt, 22nd April 2002.

SOUTHERN ITALY

Numerous different bee orchids, including a few endemics, grow in the southern part of mainland Italy. We ourselves have visited Campania in the western part of the region and southeastern Puglia (the heel of the Italian "boot"). These idyllic and luxuriant landscapes are rich in first-rate *Ophrys* habitats such as garrigue, olive groves, and open oak woods (FIG. 117). Monte Gargano in northwestern Puglia is treated separately above. In southeastern Puglia, the two areas described below are of especial interest.

South of the large ports of Bari and Brindisi, at 200-400 m altitude, there is a limestone plateau called Le Murge. On this plateau, around the cities of Alberobello, Martina Franca, Noci and Ostuni, a number of South Italian specialities occur in a most charming scenery of garrigue and open oak groves. The *Ophrys* flora includes, for example, *O. fuciflora* subsp. *apulica*, subsp. *oxyrrhynchos* and subsp. *parvimaculata*, as well as *O. bertolonii* and *O. ×flavicans*.

Large tracts of the coast line of southern Puglia have been spoiled by urbanisation, although areas of garrigue and open pine forest with plenty of bee orchids still exist east of Lecce. Southeast of San Cataldo you may encounter the only occurrences of *O. fuciflora* subsp. *candica* known from mainland Italy. It is often associated with *O. tenthredinifera* and *O. fuciflora* subsp. *apulica*, and these three bee orchids frequently hybridise in the region.

In Campania, the national park encompassing the Cilento area and Monte Alburni is a region offering extraordinary opportunities. Easily found bee orchids include *O. argolica* subsp. *biscutella*, *O. apifera*, *O. fuciflora* subsp. *fuciflora* and the beautiful *O. fuciflora* subsp. *lacaitae*. From late May to late June, variable, late-flowering forms of *O. fuciflora* can be observed. In a stony meadow north of the city of Campora we were lucky to make the first documented find of *O. fuciflora* subsp. *oxyrrhynchos* in Campania.

REFERENCES: Andrew (2002), Bernardo & Pontillo (2002), Faurholdt (2003), Galiani (1990), Gölz & Reinhard (1982).

ISTRIA

Istria is a peninsula on the eastern Adriatic coast. Nearly all of the peninsula belongs to Croatia, only the northernmost part being Slovenian. Accommodation is available along the west coast, for instance at Poreč, Pula or Rovinj. Alternatively, you can do as we did and stay at Hotel Kastel in the small picturesque medieval city of Motovun in the centre of the peninsula. Wherever you stay, the distances between accommodation and fine *Ophrys* localities in Istria are short.

Ophrys apifera and *O. sphegodes* subsp. *atrata* are widespread and common, but other *Ophrys* species such as *O. bombyliflora*, *O. fusca* and *O. bertolonii*, can only be found in the far south, where the Premantura peninsula is a superior site. The first known Istrian individual of *O. tenthredinifera* was also recently found near Pula.

One of the most frequently seen bee orchids is *O. fuciflora*, which occurs in a bewildering array of forms. Those that flower in April and May belong to subsp. *fuciflora* whereas those starting to flower in June seem referable to subsp. *elatior*. The late-flowering forms can be found, for example, in the area around Draguc, Cerovlje and Paz which is also a good site for *O. sphegodes* subsp. *sphegodes* and *O. insectifera* (FIG. 118).

In the surroundings of Kavran in the southeasternmost part of Istria, the main attraction is some colonies of highly variable bee orchids. The plants are morphologically intermediary between *O. fuciflora* and *O. scolopax* subsp. *cornuta* and belong to the partly stabilised hybrid complex, *O.* ×*vicina* nm. "*holubyana*". The same area holds fine populations of *O. sphegodes* subsp. *litigiosa*.

REFERENCES: Biel (2001), Hertel & Hertel (2002), Kerschbaumsteiner et al. (2002), Vöth & Löschl (1978).

FIG. 118. Along the idyllic road between Cerovlje and Draguc in Istria, the late-flowering *Ophrys fuciflora* subsp. *elatior* arouses interest. Photo by N. Faurholdt, 11th June 2006.

FIG. 119. From Imettos there is a magnificent view of Athens. Garrigue and cypress groves on this mountain accommodate several bee orchids, including the rare *Ophrys lutea* subsp. *melena* and *O. sphegodes* subsp. *aesculapii*. Photo by N. Faurholdt, 30th March 1998.

IMETTOS, NEAR ATHENS

A holiday in Athens, or perhaps just half a day's waiting time between two flights in the airport of the Greek capital, can be used for an excursion to Imettos on the outskirts of town. In marked contrast to the roar of polluting traffic in the city centre, peace and quiet reign in the bright air of Imettos (FIG. 119). On top of this, the hill accommodates several bee orchids, including some rare Greek specialities.

You can either take a taxi from the airport or a bus number 224 from the university in the centre of Athens to the large cemetery at Kaisariani. Do not miss the hike (c. 3 km) from the cemetery to the Kaisariani monastery further uphill as along the narrow road you will find swarms of *Ophrys* growing in open pine forest, cypress groves and garrigue (FIG. 14). *Ophrys lutea* (including subsp. *melena*) and *O. fusca* subsp. *fusca* as well as rich populations of *O. ferrum-equinum* subsp. *ferrum-equinum*, *O. tenthredinifera* and *O. speculum* subsp. *speculum* are common. The main attractions are *O. sphegodes* subsp. *aesculapii* (with or without a broad, yellow lip margin) and

O. umbilicata subsp. *umbilicata* with green sepals.

Above the monastery itself, and still accompanied by bee orchids, you can admire a magnificent view of Athens.

NAXOS

In mid-April, prior to the start of the real tourist season, Naxos is a peaceful place with a genuine Greek island atmosphere. It is a small island, the roads are narrow (FIG. 17) with only a modest amount of traffic, and Chora (the main city) is devoid of big hotels. Naxos might well be the most enchanting place where we have ever studied *Ophrys*.

Charter tours to Naxos do not start until May when most of the bee orchids have finished flowering. A trip during the peak season of *Ophrys*, therefore, requires a scheduled flight to Athens and a domestic flight onwards to Naxos which unfortunately makes the trip rather expensive.

Bee orchids grow virtually everywhere on Naxos, and all parts of the island are within easy reach from Chora. In addition to the hills surrounding Koronos (FIG. 120), we would like to draw particular attention to the area between Chora and Halkio and to the area between Halkio, Filoti and

FIG. 120. Stone walls divide the extensive grazing areas east of Koronos on Naxos into a mosaic of patches. Even in this apparently barren landscape, a number of bee orchids thrive, including the narrow endemic *Ophrys fuciflora* subsp. *andria*. Photo by N. Faurholdt, 18th April 2000.

Moni. The speciality of the island is *O. fuciflora* subsp. *andria*, which is frequent around Halkio (FIG. 15) and less so around Koronos. Scattered on Naxos you will find many populations of *O. kotschyi* subsp. *ariadnae*, whereas subsp. *cretica* seems restricted to higher altitudes on limestone hills between Galanado and Kato (east of Chora). The handsome, large-flowered *O. ×vicina* nm. *"calypsus"* can be seen at many places together with, for example, *O. scolopax* (subsp. *scolopax*, subsp. *cornuta*, subsp. *heldreichii*), *O. lutea*, *O. reinholdii* and *O. sphegodes* subsp. *mammosa*. On the northern part of the island, *O. omegaifera* subsp. *omegaifera* and *O. ferrum-equinum* subsp. *ferrum-equinum* are locally common. *Ophrys fusca* subsp. *blitopertha* and *O. sphegodes* subsp. *gortynia*, on the other hand, must be sought out in the southern part; by mid-April the former has almost finished flowering, whereas the flowers of the latter are opening.

REFERENCES: Kretzschmar & Kretzschmar (1996), Paulus & Gack (1992).

CRETE

The charter tours to this island – an absolute must for the *Ophrys* enthusiast – start in March. No matter where in Crete you choose to stay, localities rich in bee orchids will be within easy reach. We have stayed near Chania in the west, by Malia in the east, and in idyllic Agia Galini on the south coast, and on several occasions we have found *Ophrys* species just a few minutes stroll from a tourist hotel. Throughout Crete you will encounter *O. lutea* subsp. *galilaea*, *O. bombyliflora* and *O. fusca* (including subsp. *cinereophila*), and species like *O. tenthredinifera* and *O. fuciflora* are also fairly frequent.

It is difficult to rank one Cretan locality against another, but to us the area between Spili and Agia Galini (FIG. 121) is most outstanding, together with the low limestone hills near Festos and Agia Triada. Throughout these areas, a considerable number of bee orchids are common, such as *O. scolopax* subsp. *heldreichii*, *O. kotschyi* (subsp. *cretica* and subsp. *ariadnae*), *O. omegaifera* subsp. *omegaifera*, *O. sphegodes* (subsp. *cretensis* and subsp. *mammosa*) and *O. fusca* subsp. *iricolor*. In the tract between Paleochora and Kandanos in southern Crete we have seen large populations of *O. omegaifera* subsp. *omegaifera* (var. *omegaifera* and var. *basilissa*), often associated with *O. omegaifera* subsp. *fleischmannii* and fine individuals of *O. sphegodes* subsp. *spruneri*.

Bee orchids can also be found on the higher mountains. For instance, we have great *Ophrys* memories from the range surrounding the Lasithi tableland southeast of Kasteli and from the Thripti mountains northeast of Kato Chorio. In the latter area especially, large populations of O. ×*brigittae* can be encountered.

In eastern Crete, really species-rich localities are more scattered, but it may pay to search small patches of garrigue on the mountains. We have found plenty of bee orchids between Kato Chorio and Kalamafka, between Kalamafka and Anatoli, in the surroundings of Agia Vavara and Gonies and, not least, in garrigue at the seaside by Malia. Among other bee orchids, small-flowered plants of O. *kotschyi* subsp. *cretica* are to be found in eastern Crete.

REFERENCES: Alibertis (1997), Alibertis & Alibertis (1989), Henke (1986), Hölzinger & Hölzinger (1986), Kretzschmar et al. (2002), Kreutz (1990), Wellinghausen & Koch (1989).

RHODES

Nearly all hotels in Rhodes are situated in the main city at the northern tip of the island (FIG. 111A). However, it is also possible to stay at Lindos on the southeastern coast. No matter where you stay, any corner of this relatively small island can be reached in a single day.

The main attraction in the *Ophrys* flora is the fairly frequent O. *scolopax* subsp. *rhodia*, which seems endemic to Rhodes and Karpathos. Other interesting bee orchids include O. *speculum* subsp. *regis-ferdinandii* and O. *argolica* subsp. *lucis*. The former grows scattered throughout most of the island, whereas the latter is comparatively rare (personally, we have seen it north of Laerma and near Agios Issidoros south of Attaviros). North of Attaviros, the highest mountain of Rhodes (FIG. 16), there is another mountain – the wooded and botanically well-known Profitis Ilias. Here, you will be confronted by bewildering swarms of highly variable O. ×*vicina* nm. "*heterochila*" and have a chance to admire beautiful populations of O. *reinholdii*. *Ophrys omegaifera* subsp. *omegaifera* and O. *ferrum-equinum* subsp. *ferrum-equinum* occur more locally on Profitis Ilias.

Far south, by the Plimiri beach and on the plains of garrigue south of Katavia, large-flowered and small-flowered forms of O. *scolopax* subsp. *cornuta* occur together with O. *speculum* (subsp. *speculum*, subsp. *regis-ferdinandii* and hybrids). A particularly interesting bee orchid from Plimiri is O. *fusca* subsp. *blitopertha*, the only bee orchid to be consistently pollinated by a beetle.

Fine populations of O. *scolopax* subsp. *heldreichii* and large-flowered forms of O. *fuciflora* subsp. *fuciflora* are frequent in Rhodes (FIG. 111B), and the partly stabilised hybrid complex between these entities, O. ×*vicina* nm.

Fig. 121. Nea Kria Vrisi, between Spili and Agia Galini, is situated in the middle of one of the most *Ophrys*-rich areas of Crete; the orchids in the foreground are *Orchis italica* and *Serapias vomeracea* subsp. *laxiflora*. Photo by H. Æ. Pedersen, 15th April 2003.

"*calypsus*", is seen at several sites. *Ophrys fuciflora* subsp. *candica*, on the other hand, is rare on Rhodes, where we have only found it close to the Attaviros mountain.

REFERENCES: Kajan (1984), Kalteisen & Willing (1981), Kemmer & Kemmer (1987), Kretzschmar et al. (1984, 2001), Kreutz (2002), Manuel (1989, 1989a), Peter (1989), Röttger (1990), Schot (1997).

SAMOS

To get to this island in April (FIG. 122), one has to take a scheduled flight to Athens and a domestic flight onwards to Samos. If you do not like big hotels, there are plenty of opportunities for staying at lodgings or pensions, particularly in Pythagorion and Vathi in the eastern part of the island. In the western part, which holds some very interesting orchid sites, we have stayed

in a small flat in the idyllic fishing village of Ormos Marathokambos.

In March and April, bee orchids flower virtually everywhere on the island. The most frequent are *O. fusca* and *O. lutea* subsp. *galilaea*, but also *O. sphegodes* subsp. *mammosa* and *O. umbilicata* subsp. *umbilicata* are common in suitable habitats. On a peninsula east of Vathi, close to the easternmost point of Samos, you will find a monastery, Moni Zoodochu, surrounded by garrigue and open pine woods. This area accommodates an interesting *Ophrys* flora that includes, among others, *O. ×vicina* nm. "*heterochila*", *O. ferrum-equinum* subsp. *ferrum-equinum*, *O. fusca* subsp. *iricolor* and the very small-flowered *O. scolopax* var. *minutula*. A few specimens of *O. bombyliflora*, a rare species on Samos, also grow here. At 5–10 km west of Ormos Marathokampos in the southwest, there are some large and apparently barren and uninteresting, stony olive groves (FIG. 123). However, you should certainly pull up and search for three of the rarest bee orchids on

FIG. 122. A gathering thunderstorm near Mavratzeoi on central Samos; the Mediterranean weather is not consistently hospitable in spring, and one should always be prepared for a few cold and rainy days. Photo by H. Æ. Pedersen, 13th April 1999.

Fig. 123. Less than 10 km west of the small fishing hamlet of Ormos Marathokampos on Samos a surprisingly high number of *Ophrys* species can be found in extensive olive groves on stony and apparently barren ground. Photo by N. Faurholdt, 8th April 1998.

Samos, *O. omegaifera* subsp. *omegaifera, O. speculum* subsp. *regis-ferdinandii* and *O. fusca* subsp. *blitopertha,* all of which occur in this area. Additional bee orchids to be found in these olive groves include, for example, *O.* ×*vicina* nm. "*heterochila*" and *O. scolopax* var. *minutula.*

Ophrys fuciflora subsp. *fuciflora* and *O. speculum* subsp. *speculum* also occur, scattered throughout the island.

REFERENCES: Anderson & Anderson (1989), Gölz & Reinhard (1981), Hirth & Spaeth (1989, 1992).

LESBOS

As with Naxos and Samos, charter tours to Lesbos do not start until the main flowering season of *Ophrys* has passed, so a trip to this island in March or April involves a scheduled flight to Athens and a domestic flight onwards. We have stayed at a hotel in Mitilini (the main city), which is situated in the southeast and constitutes an excellent base for *Ophrys* excursions. The most interesting localities are found partly in the southeastern part of the island, partly north of Antissa in the northwest.

Each island in the Aegean usually holds one or more bee orchids that are particularly common on that island (or perhaps nearly endemic), and, of course, one is especially interested to find these particular plants. On Lesbos, special bee orchids include *O. umbilicata* subsp. *bucephala* and *O. argolica* subsp. *lesbis* (and to a lesser extent *O. scolopax* var. *minutula*). On the southern part of the island, around Agios Isidores, Plomari and Megalochori, there are good chances of finding *O. umbilicata* subsp. *bucephala* on roadside slopes and in old olive groves (FIG. 124). *Ophrys argolica* subsp. *lesbis*, which was long considered endemic to Lesbos, can be found in garrigue between Antissa and Gravathas in the northeast. *Ophrys scolopax* var. *minutula* is fairly frequent in the southeast; in stony olive groves west of Moria (FIG. 19) it grows in profusion with other bee orchids such as *O. speculum* subsp. *speculum*, *O. ferrum-equinum* subsp. *ferrum-equinum*, *O. umbilicata* subsp. *umbilicata* and *O. fusca* subsp. *iricolor*.

The most frequently occurring bee orchids on Lesbos are *O. fusca* subsp. *fusca*, *O. lutea* subsp. *galilaea* and *O. speculum* subsp. *speculum*, whereas *O. tenthredinifera*, *O. reinholdii* and *O. fusca* subsp. *blitopertha* are more or less rare.

REFERENCES: Biel (1998, 1999), Gölz & Reinhard (1981, 1989).

FIG. 124. This olive grove west of Lambou Milli on Lesbos held a wide range of flowering orchids, including bee orchids, until a shepherd drove his sheep across the area; after that, not a single inflorescence could be retrieved! Photo by N. Faurholdt, 1st April 1996.

Bibliography and references

Asterisks indicate the publications that we took into account when compiling the synonymies in the systematic account (only names accepted in the publications were considered).

Ackermann, M. & Ackermann, M. 1988. Orchideenfunde rund um den Ätna. *Mitt. Arbeitskreis Heimische Orchid. Baden-Württemberg* 20: 805-816.

Ackermann, M. & Ackermann, M. 1992. Einige Orchideenfunde aus Südspanien. *Ber. Arbeitskreis. Heimische Orchid.* 9(2): 64-69.

Ågren, L. & Borg-Karlson, A.-K. 1984. Responses of *Argogorytes* (Hymenoptera: Sphecidae) males to odor signals from *Ophrys insectifera* (Orchidaceae). Preliminary EAG and chemical investigation. *Nov. Acta Reg. Soc. Sci. Ups., Ser. V-C* 3: 111-117.

Ågren, L., Kullenberg, B. & Sensenbaugh, T. 1984. Congruences in pilosity between three species of *Ophrys* (Orchidaceae) and their hymenopteran pollinators. *Nov. Acta Reg. Soc. Sci. Ups., Ser. V-C* 3: 15-25.

Alibertis, A. 1997. *Die Orchideen von Kreta und Karpathos*. Antonis Alibertis, Iraklion.

Alibertis, C. & Alibertis, A. 1989. *The wild orchids of Crete*. 2nd ed. C. & A. Alibertis, Iraklion.

Alibertis, C., Alibertis, A. & Reinhard, H. R. 1990. Untersuchungen am *Ophrys omegaifera*-Komplex Kretas. *Mitt. Arbeitskreis Heimische Orchid. Baden-Württemberg* 22: 181-236.

Anderson, B. & Anderson, E. 1989. The wild orchids of Samos. *Mitt. Arbeitskreis Heimische Orchid. Baden-Württemberg* 21: 1136-1155.

Andrew, S. 2002. Fabulous Apulia, land of the *Ophrys*. *Hardy Orchid Soc. Newslett.* 23: 12-18.

Arnold, M. L. 1997. *Natural hybridization and evolution*. Oxford University Press, New York & Oxford.

Ascensão, L., Francisco, A., Cotrim, H. & Salomé Pais, M. 2005. Comparative structure of the labellum in *Ophrys fusca* and *O. lutea* (Orchidaceae). *Am. J. Bot.* 92: 1059-1067.

Ashley, J. M., Gintzburger, G. A. A. & Hossain, E. 1982. *Ophrys lutea* (Gouan) Cav. var. *minor* Guss. and *Ophrys fusca* Link subsp. *fusca* Nels. – new records for Libya. *Amer. Orchid Soc. Bull.* 51: 15-20.

Ayasse, M., Schiestl, F. P., Paulus, H. F., Ibarra, F. & Francke, W. 2003. Pollinator attraction in a sexually deceptive orchid by means of unconventional chemicals. *Proc. Roy. Soc. London., Ser. B, Biol. Sci.* 270: 517-522.

Ayasse, M., Schiestl, F. P., Paulus, H. F., Löfstedt, C., Hansson, B., Ibarra, F. & Francke, W. 2000. Evolution of reproductive strategies in the sexually deceptive orchid *Ophrys sphegodes*: how does flower-specific variation of odor signals influence reproductive success? *Evolution* 54: 1995-2006.

Bartolo, G., Lanfranco, E., Pulvirenti, S. & Stevens, D. T. 2001. Le Orchidaceae dell'arcipelago maltese (Mediterraneo centrale). *J. Eur. Orchid.* 33: 743-870.

Bartolo, G. & Pulvirenti, S. 1997. *Ophrys calliantha* (Orchidaceae): una nuova specie della Sicilia. *Caesiana* 9: 41-47.

Bartolo, G. & Pulvirenti, S. 1997a. A check-list of Sicilian orchids. *Bocconea* 5: 797-824.

Bartolo, G. & Pulvirenti, S. 2005. Le orchidee della Sicilia: aggiornamento della check-list. *J. Eur. Orchid.* 37: 585–623.

Bateman, R. M. 1999. Integrating molecular and morphological evidence of evolutionary radiations. In: Hollingsworth, P. M., Bateman, R. M. & Gornall, R. J. (eds), *Molecular systematics and plant evolution.* Taylor & Francis, London & New York, pp. 432–471.

Bateman, R. M. 2001. Evolution and classification of European orchids: insights from molecular and morphological characters. *J. Eur. Orchid.* 33: 33–119.

Bateman, R. M. & DiMichele, W. A. 2002. Generating and filtering major phenotypic novelties: neoGoldschmidtian saltation revisited. In: Cronk, Q. C. B., Bateman, R. M. & Hawkins, J. A. (eds), *Developmental genetics and plant evolution.* Taylor & Francis, London & New York, pp. 109–159.

Bateman, R. M., Hollingsworth, P. M., Preston, J., Luo, Y.-B., Pridgeon, A. M. & Chase, M. W. 2003. Molecular phylogenetics and evolution of Orchidinae and selected Habenariinae (Orchidaceae). *Bot. J. Linn. Soc.* 142: 1–40.

Bateman, R. M. & Rudall, P. J. 2006. The good, the bad, and the ugly: using naturally occurring terata to distinguish the possible from the impossible in orchid floral evolution. *Aliso* 22: 481–496.

Baum, A., Baum, H., Claessens, J. & Kleynen, J. 2002. *Ophrys apifera* Hudson, eine variable Art. *Jahresber. Naturwiss. Vereins Wuppertal* 55: 78–94.

Baumann, B. & Baumann, H. 1990. Zur Höhenverbreitung der Gattung *Ophrys* L. *Mitt. Arbeitskreis Heimische Orchid. Baden-Württemberg* 22: 818–829.

Baumann, H. 1975. Die *Ophrys*-Arten der Sektion *Fusci-Luteae* Nelson in Nordafrika. *Orchidee (Hamburg)* 26: 132–140.

Baumann, H. & Dafni, A. 1981. Differenzierung und Arealform des *Ophrys omegaifera*-Komplexes im Mittelmeergebiet. *Beih. Veröff. Naturschutz Landschaftspflege Baden-Württemberg* 19: 129–153.

Baumann, H., Giotta, C., Künkele, S., Lorenz, R. & Piccitto, M. 1995. *Ophrys holoserica* subsp. *chestermanii* J. J. Wood eine gefährdete und endemische Orchidee von Sardinien. *J. Eur. Orchid.* 27: 185–244.

★Baumann, H. & Künkele, S. 1982. *Die wildwachsenden Orchideen Europas.* Franckh'sche Verlagshandlung, Stuttgart.

Baumann, H. & Künkele, S. 1984. Über *Ophrys exaltata* Ten. und *Ophrys crabronifera* Mauri. *Mitt. Arbeitskreis Heimische Orchid. Baden-Württemberg* 16: 633–663.

★Baumann, H. & Künkele, S. 1986. Die Gattung *Ophrys* L. – eine taxonomische Übersicht. *Mitt. Arbeitskreis Heimische Orchid. Baden-Württemberg* 18: 305–688.

★Baumann, H. & Künkele, S. 1988. *Die Orchideen Europas. 247 Arten, 51 Unterarten.* Franckh'sche Verlagshandlung, W. Keller & Co., Stuttgart.

Baumann, H. & Künkele, S. 1995. *Ophrys argolica* H. Fleischm. Orchidaceae. Vulnerable. In: Phitos, D., Strid, A., Snogerup, S. & Greuter, W. (eds), *The red data book of rare and threatened plants of Greece.* World Wide Fund for Nature (WWF), Athens, pp. 382–383.

Baumann, H. & Künkele, S. 1995a. *Ophrys umbilicata* Desf. subsp. *rhodia* H. Baumann & Künkele. Orchidaceae. Rare. In: Phitos, D., Strid, A., Snogerup, S. & Greuter, W. (eds), *The red data book of rare and threatened plants of Greece.* World Wide Fund for Nature (WWF), Athens, pp. 386–387.

Baumann, H. & Künkele, S. 1995b. *Ophrys helenae.* Orchidaceae. Rare. In: Phitos, D., Strid, A., Snogerup, S. & Greuter, W. (eds), *The red data book of rare and threatened plants of Greece*. World Wide Fund for Nature (WWF), Athens, pp. 384–385.

Baumbach, N. 2003. *Ophrys tenthredinifera* Willd. in Istrien. *Orchidee (Hamburg)* 54: 674.

Beniston, N. T. & Beniston, W. S. 1999. *Wild Orchids of Mallorca*. Editorial Moll, Mallorca.

Berger, L. 2003. Observations sur le comportement de quelques pollinisateurs d'orchidées (2e partie). *Orchidophile* 159: 277–290.

Bergman, B., Draleva, S. & Uzunov, S. 2004. *Ophrys reinholdii* (Orchidaceae) – a new species for the Bulgarian flora. *Phytol. Balcan.* 10: 175–177.

Bernardo, L. & Pontillo, D. 2002. *Le orchidee spontanee della Calabria*. Edizioni Prometeo, Castrovillari.

Bernardos, S., Amich, F. & Gallego, F. 2003. Karyological and taxonomic notes on *Ophrys* (Orchidoideae, Orchidaceae) from the Iberian Peninsula. *Bot. J. Linn. Soc.* 142: 395–406.

Bernardos, S., Crespi, A., del Rey, F. & Amich, F. 2005. The section *Pseudophrys* (*Ophrys*, Orchidaceae) in the Iberian Peninsula: a morphometric and molecular analysis. *Bot. J. Linn. Soc.* 148: 359–375.

Biel, B. 1998. Die Orchideenflora der Insel Lesvos (Griechenland). *J. Eur. Orchid.* 30: 251–443.

Biel, B. 1999. Nachtrag zur Orchideenflora von Lesvos (Griechenland). *J. Eur. Orchid.* 31: 852–876.

Biel, B. 2001. Zwei Exkursionen der AHO Unterfranken zur Halbinsel Istrien (Kroatien). *Ber. Arbeitskreis. Heimische Orchid.* 18(1): 133–161.

Biel, B. 2002. Zur Systematik und Nomenklatur der europäischen Orchideen – unter Bezug auf aktuelle Veröffentlichungen. *Ber. Arbeitskreis. Heim. Orchid.* 19(1): 44–52.

Bjørndalen, J. E. 2006. *Ophrys insectifera* at the edge of its geographical range: aspects of ecology, vegetational affiliation and conservation in Norway. *J. Eur Orchid.* 38: 415–448.

Blanco, M. A. & Barboza, G. 2005. Pseudocopulatory pollination in *Lepanthes* (Orchidaceae: Pleurothallidinae) by fungus gnats. *Ann. Bot.* 95: 763–772.

Blatt, H. & Hertel, H. 1984. Beitrag zur Verbreitung der Orchideen auf den Balearen. *Ber. Arbeitskreis. Heimische Orchid.* 1: 41–70.

Blatt, H. & Wirth, W. 1990. Anmerkungen zu "Die nomenklatorischen Typen der von Linnaeus veröffentlichten Namen europäischer Orchideen". I *Ophrys fuciflora* versus *Ophrys holoserica*. *Ber. Arbeitskreis. Heimische Orchid.* 7(1): 4–8.

Bockhacker, K. 1996. Beobachtungen zur Populationsentwicklung und zum Blühverhalten der Bienen-Ragwurz. *Ber. Arbeitskreis. Heimische Orchid.* 12(2): 34–47.

Borg-Karlson, A.-K. 1990. Chemical and ethological studies of pollination in the genus *Ophrys* (Orchidaceae). *Phytochemistry* 29: 1359–1387.

★Bournérias, M. (ed.) 1998. *Les orchidées de France, Belgique et Luxembourg*. Collection Parthénope & Société Francaise d'Orchidophilie, Paris.

Bremer, K. & Janssen, T. 2006. Gondwanan origin of major monocot groups inferred from dispersal-vicariance analysis. In: Columbus, J. T., Friar, E. A., Porter, J. M., Prince, L. M. & Simpson, M. G. (eds), Monocots: comparative biology and evolution (excluding Poales). *Aliso* 22: 22–27 (i-x, 1–735).

Büel, H. 1978. Beobachtungen über die Bestäubung von *Ophrys bertolonii* Mor. *Orchidee (Hamburg)* 29: 106–109.

Buttler, K. P. 1983. Die *Ophrys-ciliata* (*speculum*)-Gruppe, eine Neubewertung (Orchidaceae: Orchideae). In: Senghas, K. & Sundermann, H. (eds), Probleme der

Taxonomie, Verbreitung und Vermehrung europäischer und mediterraner Orchideen. *Orchidee (Hamburg)*, special issue March 1983: 37-57 (1-120).

★Buttler, K. P. 1986. *Orchideen. Die wildwachsenden Arten und Unterarten Europas, Vorderasiens und Nordafrikas.* Mosaik Verlag Gmbh., Munich.

Caporali, E., Grünanger, P., Marziani, G., Servettaz, O. & Spada, A. 2001. Molecular (RAPD) analysis of some taxa of the *Ophrys bertolonii* aggregate (Orchidaceae). *Israel J. Pl. Sci.* 49: 85-89.

Chase, M. W. 2005. Classification of Orchidaceae in the age of DNA data. *Curtis's Bot. Mag.* 22: 2-7 & 1 pl.

van der Cingel, N. A. 1995. *An atlas of orchid pollination. European orchids.* A. A. Balkema, Rotterdam.

van der Cingel, N. A. 2001. *An atlas of orchid pollination. America, Africa, Asia and Australia.* A. A. Balkema, Rotterdam & Brookfield.

Claessens, J. 1992. Einige opmerkingen over de orchideeën van Andalusië (Spanje). *Eurorchis* 4: 37-52.

Claessens, J. & Kleynen, J. 2002. Investigations on the autogamy in *Ophrys apifera* Hudson. *Jahresber. Naturwiss. Vereins Wuppertal* 55: 62-77.

Cribb, P. & Govaerts, R. 2005. Just how many orchids are there? In: Raynal-Roques, A. & Roguenant, A. (eds.), *Actes du 18e congrès mondial et exposition d'orchidées 11-20 Mars 2005, Dijon-France.* Naturalia Publications, Turriers, pp. 161-172.

Dafni, A. & Bernhardt, P. 1990. Pollination of terrestrial orchids of southern Australia and the Mediterranean region: systematic, ecological, and evolutionary implications. *Evol. Biol.* 24: 193-252.

★Danesch, E. & Danesch, O. 1969. *Orchideen Europas. Südeuropa.* Verlag Hallwag, Bern & Stuttgart.

Danesch, E. & Danesch, O. 1972. *Orchideen Europas. Ophrys-Hybriden.* Hallwag Verlag, Bern & Stuttgart.

Danesch, O. & Danesch, E. 1964. Orchideen der Gattung *Ophrys* auf Mallorca. In: Sundermann, H. & Haber, W. (eds), Probleme der Orchideengattung *Ophrys. Orchidee (Hamburg)*, special issue March 1964: 34-35 (1-72 & 6 pls).

Danesch, O. & Danesch, E. 1971. *Ophrys bertoloniiformis* O. & E. Danesch, sp. nov., eine Sippe hybridogenen Ursprungs. *Orchidee (Hamburg)* 22: 115-117.

Danesch, O. & Danesch, E. 1971a. *Ophrys promontorii* O. et E. Danesch sp. nov., eine hybridogene Sippe aus Süditalien. *Orchidee (Hamburg)* 22: 256-258.

Danesch, O. & Danesch, E. 1976. Hybriden und Hybridenschwärme aus *Ophrys argolica* Fleischm. und *Ophrys scolopax* Cav. ssp. *cornuta* (Stev.) E. G. Cam. in Griechenland. In: Senghas, K. (ed.), *Tagungsbericht der 8. Welt-Orchideen-Konferenz Palmengarten Frankfurt 10.-17. April 1975.* Deutsche Orchideen Gesellschaft e.V., Frankfurt am Main, pp. 129-138.

★Davies, P., Davies, J. & Huxley, A. 1983. *Wild orchids of Britain and Europe.* The Hogarth Press, London.

Del Prete, C. 1984. The genus "*Ophrys*" L. (Orchidaceae): a new taxonomic approach. *Webbia* 38: 209-220.

Del Prete, C. 1999: The OPTIMA project for mapping Mediterranean orchids: the situation in Italy and a provisional checklist. In: Uotila, P. (ed.), Chorological problems in the European flora. *Proceedings of the VIII Meeting of the Committee for Mapping the Flora of Europe. Helsinki, Finland, 8-10 August 1997. Acta Bot. Fenn.* 162: 145-154 (i-xiv, 1-196).

Del Prete, C., Tichy, H. & Tosi, G. 1993. *Le orchidee spontanee della Maremma Grossetana*. PRO.GRA.MS Italia, Libr. Massimi, Porto Ercole.

Del Prete, C. & Tosi, G. 1981. *Orchidee Spontanee dell'Argentario*. Pitigliano, Cagliari.

★Del Prete, C. & Tosi, G. 1988. *Orchidee spontanee d'Italia*. Monografia e iconografia. U. Mursia editore S.P.A., Milano.

Delforge, P. 1989. Le groupe d'*Ophrys bertolonii* Moretti. *Mém. Soc. Roy. Bot. Belg.* 11: 7-29.

★Delforge, P. 1994. *Guide des orchidées d'Europe, d'Afrique du Nord et du Proche-Orient*. Delachaux et Niestlé S.A., Lausanne.

Delforge, P. 1996. Europe, North Africa, and the Near East. In: Hágsater, E. & Dumont, V. (eds), *Orchids. Status survey and conservation action plan*. IUCN, Gland & Cambridge.

★Delforge, P. 2001. *Guide des orchidées d'Europe, d'Afrique du Nord et du Proche-Orient*. 2nd ed. Delachaux et Niestlé S.A., Lausanne.

★Delforge, P. 2005. *Guide des orchidées d'Europe, d'Afrique du Nord et du Proche-Orient*. 3rd ed. Delachaux et Niestlé, Paris.

D'Emerico, S., Pignone, D., Bartolo, G., Pulvirenti, S., Terrasi, C., Stuto, S. & Scrugli, A. 2005. Karyomorphology, heterochromatin patterns and evolution in the genus *Ophrys* (Orchidaceae). *Bot. J. Linn. Soc.* 148: 87-99.

Devillers, P. & Devillers-Terschuren, J. 1994. Essai d'analyse systematique du genre *Ophrys. Naturalistes Belges* 75, suppl.: 273-400.

Devillers, P. & Devillers-Terschuren, J. 2000. Observation sur les ophrys du groupe d'*Ophrys subfusca* en Tunisie. *Naturalistes Belges* 81: 283-297.

Devillers, P., Devillers-Terschuren, J. & Tyteca, D. 2003. Notes on some of the taxa comprising the group of *Ophrys tenthredinifera* Willdenow. *J. Eur. Orchid.* 35: 109-161.

Dimitrov, D., Gussev, C., Denchev, C., Koeva, Y. & Pavlova, D. 2001. *Ophrys argolica* (Orchidaceae), a new species to the Bulgarian flora. *Phytologia Balcan.* 7: 199-200.

Dorland, E. & Willems, J. H. 2002. Light climate and plant performance of *Ophrys insectifera*; a four-year field experiment in The Netherlands (1998-2001). In: Kindlmann, P., Willems, J. H. & Whigham, D. F. (eds), *Trends and fluctuations and underlying mechanisms in terrestrial orchid populations*. Backhuys Publishers, Leiden, pp. 225-238.

Dorland, E. & Willems, J. H. 2006. High light availability alleviates the costs of reproduction in *Ophrys insectifera* (Orchidaceae). *J. Eur. Orchid.* 38: 369-386.

Dressler, R. L. 1981. *The orchids. Natural history and classification*. Harvard University Press, Cambridge Mass. & London.

Dressler, R. L. 1993. *Phylogeny and classification of the orchid family*. Cambridge University Press, Cambridge.

Eberhardt, K. 1995. *Ophrys scolopax* ssp. *apiformis*, erstmals gefunden auf Ibiza. *Ber. Arbeitskreis. Heimische Orchid.* 12(1): 77.

Edmondson, T. 1979. *Ophrys apifera* Huds. in artificial habitats. *Watsonia* 12: 337-338.

Ehrendorfer, F. 1980. Hybridisierung, Polyploidie und Evolution bei europäisch-mediterranen Orchideen. In: Senghas, K. & Sundermann, H. (eds), Probleme der Evolution bei europäischen und mediterranen Orchideen. *Orchidee (Hamburg)*, special issue November 1980: 15-34 (1-178 & 2 pls).

Engel, R. & Mathé, H. 2002. *Orchidées sauvages d'Alsace et des Vosges*. Editions du Griffon, Saverne.

Falci, A. & Giardina, S. A. 2000. Segnalazioni di Orchidaceae in Sicilia. *J. Eur. Orchid.* 32: 279-290.

Faurholdt, N. 2003. Zur Verbreitung von *Ophrys holoserica* (Burm. fil.) Greuter subsp. *oxyrrhynchos* (Tod.) H. Sundermann auf den süditalienischen Festland. *Ber. Arbeitskreis. Heimische Orchid.* 19(2): 27-32.

Faurholdt, N. 2003a. Notes on the genus *Ophrys* in Cyprus and Israel. *J. Eur. Orchid.* 35: 739-749.

Fay, M. F. & Krauss, S. L. 2003. Orchid conservation genetics in the molecular age. In: Dixon, K. W., Kell, S. P., Barrett, R. L. & Cribb, P. J. (eds), *Orchid conservation*. Natural History Publications (Borneo), Kota Kinabalu, pp. 91-112.

Frosch, W. 1983. Asymbiotische Vermehrung von *Ophrys holoserica* mit Blüten nach 22 Monaten. *Orchidee (Hamburg)* 34: 58-61.

Galesi, R. 1995. Contributo alla conoscenza delle Orchidacee del territorio di Niscemi (Sicilia) e presentazione di due nuovi ibridi. *J. Eur. Orchid.* 27: 252-284.

Galesi, R. 1999. Le Orchidaceae della Riserva Naturale Orientata "Pino d'Aleppo" (Ragusa, Sicilia meridionale). *J. Eur. Orchid.* 31: 297-328.

Galiani, G. G. 1990. *Le orchidee spontanee Europe. Un tesoro naturale della Murgia.* Gruppo Di Informazione Ambientale, Alberobello, Artigrafiche Pugliesi s.n.c., Martina Franca.

Gerasimova, I., Petrova, A. & Venkova, D. 1998. *Ophrys apifera* Hudson reestablished in the Bulgarian flora. *Phytologia Balcan.* 4: 53-55.

Giotta, C. & Piccitto, M. 1990. *Orchidee spontanee della Sardegna – guida al riconoscimento delle specie.* Carlo Delfino editore, Sassari.

Giotta, C. & Piccitto, M. 1991. Die wildwachsenden Orchideen der Ogliastra (mittleres Ostsardinien). *Mitt. Arbeitskreis Heimische Orchid. Baden-Württemberg* 23: 247-306.

Gölz, P. & Reinhard, H. R. 1975. Biostatistische Untersuchungen über *Ophrys bertoloniiformis* O. et E. Danesch. *Ber. Schweiz. Bot. Ges.* 85: 31-56.

Gölz, P. & Reinhard, H. R. 1980. Populationsstatistische Analysen bestätigen die Heterogenität von "*Ophrys arachnitiformis*" (Orchidaceae). *Pl. Syst. Evol.* 136: 7-39.

Gölz, P. & Reinhard, H. R. 1981. Die Orchideenflora der ostägäischen Inseln Kos, Samos, Chios und Leswos (Griechenland). *Beih. Veröff. Naturschutz Landespflege Baden-Württemberg* 19: 5-127.

Gölz, P. & Reinhard, H. R. 1982. Orchideen in Sueditalien. Ein Beitrag zur Kenntnis der Orchideenflora Apuliens, der Basilicata, Kalabriens und Siziliens. *Mitt. Arbeitskreis Heimische Orchid. Baden-Württemberg* 14: 1-124.

Gölz, P. & Reinhard, H. R. 1989. Zur Orchideenflora von Lesvos. *Mitt. Arbeitskreis Heimische Orchid. Baden-Württemberg* 21: 1-87 & 1 pl.

Gölz, P. & Reinhard, H. R. 1990. Beitrag zur Kenntnis der Orchideenflora Sardiniens (2. Teil). *Mitt. Arbeitskreis Heimische Orchid. Baden-Württemberg* 22: 405-510.

Gölz, P. & Reinhard, H. R. 1996. Gestaltwandel innerhalb kretischer Orchideenaggregate im Verlauf der Monate Januar bis Mai. *J. Eur. Orchid.* 28: 641-701.

Gölz, P. & Reinhard, H. R. 2001. Der ostmediterrane und anatolische *Ophrys holoserica*-Komplex – "Splitter" kontra "Lumper". *J. Eur. Orchid.* 33: 941-1024.

Grant, V. 1981. *Plant speciation.* 2nd ed. Columbia University Press, New York.

Grasso, M. P. & Grillo, M. 1996. Le Orchidaceae dell'Etna. *J. Eur. Orchid.* 28: 119-215.

Grasso, M. P. & Manca, L. 2002. Über die Bestäuber einiger Sippen aus der Gattung *Ophrys* im Sarcidano (Sardinien). *J. Eur. Orchid.* 34: 733-738.

Greilhuber, J. & Ehrendorfer, F. 1975. Chromosome numbers and evolution in *Ophrys* (Orchidaceae). *Pl. Syst. Evol.* 124: 125-138.

Greuter, W. 2004. (1645) Proposal to conserve the name *Ophrys speculum* (Orchidaceae) with a conserved type. *Taxon* 53: 1070-1071.

Grünanger, P., Caporali, E., Marziani, G., Menguzzato, E. & Servettaz, O. 1998. Molecular (RAPD) analysis on Italian taxa of the *Ophrys bertolonii* aggregate (Orchidaceae). *Pl. Syst. Evol.* 212: 177-184.

Gulli, V., Tosi, G., Filippi, L. & Del Prete, C. 2003. On the pollination of some orchids of the genus *Ophrys* at Mount Argentario (Grosseto, Central-Western Italy). II. *O. bertolonii* Moretti, *O. fuciflora* (F. W. Schmidt) Moench subsp. *fuciflora* and *O. bombyliflora* Link. *Caesiana* 20: 35-43.

Gulyás, G., Sramkó, G., Molnár, A., Rudnóy, S., Illyés, Z., Balázs, T. & Bratek, Z. 2005. Nuclear ribosomal DNA ITS paralogs as evidence of recent interspecific hybridization in the genus *Ophrys* (Orchidaceae). *Acta Biol. Cracov. Ser. Bot.* 47: 61-67.

Hágsater, E. 1996. Orchid Specialist Group. *Species* 26-27: 114-115.

Hahn, W. & Salkowski, H.-E. 2005. Zur Kenntnis von *Ophrys flavicans* Visiani. *Ber. Arbeitskreis. Heimische Orchid.* 21(1): 48-58.

Hansen, K., Hansen, R.-B., Kreutz, C. A. J., Rückbrodt, U. & Rückbrodt, D. 1990. Beitrag zur Kenntnis und Verbreitung der Orchideenflora von Zypern mit Interims-Verbreitungskarten. *Mitt. Arbeitskreis Heimische Orchid. Baden-Württemberg* 22: 73-171.

Hansson, S. 2001. *Ophrys aegaea* auf Zypern. *J. Eur. Orchid.* 33: 925.

Henke, E. 1986. Exkursionen in die Orchideenflora Kretas. *Ber. Arbeitskreis. Heimische Orchid.* 3: 13-38.

Hennecke, G. & Hennecke, M. 2000. Fundortangaben von Sardinien (20.4. bis 29.4.2000). *J. Eur. Orchid.* 32: 673-676.

Hermjakob, G. 1976. Orchideenstandorte in Südeuropa: aus der Flora Griechenlands. In: Senghas, K. (ed.), *Tagungsbericht der 8. Welt-Orchideen-Konferenz Palmengarten Frankfurt 10-17 April 1975*. Deutsche Orchideen Gesellschaft e.V., Frankfurt am Main, pp. 95-98.

Hermosilla, C. E., Amardeilh, J.-P. & Soca, R. 1999. *Sterictiphora furcata* Villers, pollinisateur d'*Ophrys subinsectifera* Hermosilla & Sabando. *Orchidophile* 139: 247-254.

Hermosilla, C. E. & Soca, R. 1999. Distribution of *Ophrys aveyronensis* (J. J. Wood) Delforge (Orchidaceae) and survey of its hybrids. *Caesiana* 13: 31-38.

Hertel, H. 1984. Beiträge zur Verbreitung der Orchideen auf Sizilien. *Ber. Arbeitskreis. Heimische Orchid.* 1: 167-174.

Hertel, H. 1989. Beiträge zur Verbreitung der Orchideen in Andalusien. *Ber. Arbeitskreis Heimische Orchid.* 6(1): 106-111.

Hertel, S. & Hertel, K. 2002. Beobachtungen zu den Orchideen Istriens. *J. Eur. Orchid.* 34: 493-542.

Hertel, S. & Hertel, K. 2003. Die Orchideen der Inseln Cres und Lošinj. *J. Eur. Orchid.* 35: 685-721.

Hervouet, C. & Hervouet, J.-M. 1998. Quelques observations sur les orchidées de Malte et de Tunisie. *Orchidophile* 130: 28-34.

Hervouet, J.-M. 1984. *Ophrys bornmuelleri* á Rhodes. *Orchidophile* 64: 710.

Hill, D. A. 1978. A seven year study of a colony of bee orchids (*Ophrys apifera*

Hudson). *Watsonia* 12: 162-163.

Hirth, M. & Spaeth, H. 1989. Die Orchideen der Insel Samos. Ein Beitrag zur Kartierung der Orchideen des Mittelmeerraumes. *Mitt. Arbeitskreis Heimische Orchid. Baden-Württemberg* 21: 1068-1135.

Hirth, M. & Spaeth, H. 1992. Zur Orchideenflora von Samos. *Mitt. Arbeitskreis Heimische Orchid. Baden-Württemberg* 24: 1-51.

Hirth, M. & Spaeth, H. 1998. Zur Orchideenflora von Chios – *Ophrys homeri*, eine neue Ophrysart. *J. Eur. Orchid.* 30: 3-80.

Hoffmann, V. 1983. Orchideenkartierung Mallorca. *Mitt. Arbeitskreis Heimische Orchid. Baden-Württemberg* 15: 109-151.

Hölzinger, C. & Hölzinger, J. 1986. Beiträge zur Orchideenflora von Kreta. *Mitt. Arbeitskreis Heimische Orchid. Baden-Württemberg* 18: 137-150.

Horsfall, A. & Wigginton, M. J. 1999. *Ophrys sphegodes* Miller (Orchidaceae). In: Wigginton, M. J. (ed.), *British red data books 1. Vascular plants.* 3rd ed. Joint Nature Conservation Committee, Peterborough, pp. 259-260.

Hutchings, M. J. 1987. The population biology of the early spider orchid, *Ophrys sphegodes* Mill. I. A demographic study from 1975 to 1984. *J. Ecol.* 75: 711-727.

Hutchings, M. J. 1987a. The population biology of the early spider orchid, *Ophrys sphegodes* Mill. II. Temporal patterns in behaviour. *J. Ecol.* 75: 729-742.

Hutchings, M. J. 1989. Population biology and conservation of *Ophrys sphegodes*. In: Pritchard, H. W. (ed.), *Modern methods in orchid conservation: the role of physiology, ecology and management.* Cambridge University Press, Cambridge, pp. 101-115.

Jacobsen, N. & Rasmussen, F. N. 1976. Über die Bestäubung von *Ophrys speculum* Link auf Mallorca. *Orchidee (Hamburg)* 27: 64-66.

Janssen, T. & Bremer, K. 2004. The age of major monocot groups inferred from 800+ *rbcL* sequences. *Bot. J. Linn. Soc.* 146: 385-398.

Johansen, B. & Frederiksen, S. 2002. Orchid flowers: evolution and molecular development. In: Cronk, Q. C. B., Bateman, R. M. & Hawkins, J. A. (eds), *Developmental genetics and plant evolution.* Taylor & Francis, London & New York, pp. 206-219.

Kajan, E. 1984. Osterurlaub 1983 auf der Sonneninsel Rhodos. *Ber. Arbeitskreis. Heimische Orchid.* 1: 71-75.

Kajan, E. 1987. Orchideenfunde auf dem italienischen Festland und auf Sizilien. *Ber. Arbeitskreis. Heimische Orchid.* 4: 131-144.

Kalteisen, M. & Reinhard, H. R. 1987. Zwei neue *Ophrys*-Taxa (Orchidaceae) aus dem Ägäischen Archipel: *Ophrys aegaea* Kalteisen & Reinhard, spec. nov., *Ophrys aegaea* subsp. *lucis* Kalteisen & Reinhard, subsp. nov. *Mitt. Arbeitskreis Heimische Orchid. Baden-Württemberg* 19: 895-938.

Kalteisen, M. & Willing, E. 1981. Verbreitungskarten der Orchideen von Rhodos. *Mitt. Arbeitskreis Heimische Orchid.* 13: 377-446.

Kapteyn den Boumeester, D. W. 1997. The establishment of the "Secretariat for the Conservation of European Orchids". *Eurorchis* 9: 83-86.

Keitel, C. & Remm, W. 1991. Die Orchideenflora der Inseln Simi, Tilos und Nisyros. *Mitt. Arbeitskreis Heimische Orchid. Baden-Württemberg* 23: 81-106.

Kell, S. P., Pradhan, U. C., Seaton, P. T. & Cribb, P. J. 2003. The IUCN/SSC Orchid Specialist Group: successes and future challenges. In: Dixon, K. W., Kell, S. P., Barrett, R. L. & Cribb, P. J. (eds), *Orchid conservation.* Natural History Publications (Borneo), Kota Kinabalu, pp. 299-312.

Kemmer, U. & Kemmer, A. 1987. Orchideensuche auf Rhodos. *Mitt. Arbeitskreis Heimische Orchid. Baden-Württemberg* 19: 853-865.

Kerschbaumsteiner, H., Perko, M. L. & Stimpfl, G. 2002. Die Orchideenflora Istriens und der Kvarner Inseln Krk, Cres und Lošinj – ein Vorbericht der Arbeitsgruppe. *J. Eur. Orchid.* 34: 115-127.

Klein, E. 1978. Hyperchrome und apochrome Orchideenblüten. *Orchidee (Hamburg)* 29: 21-31.

Klinger, P. U. 1974. Zur Orchideenflora von Ulassai/Ostsardinien. *Orchidee (Hamburg)* 25: 218-222.

Koopowitz, H. 2001. *Orchids and their conservation*. B. T. Batsford Ltd., London.

Krämer, E. & Krämer, K. 1992. Orchideenkartierung Mallorca. *Ber. Arbeitskreis. Heimische Orchid.* 9(2): 70-86, 195.

Krämer, E. & Krämer, K. 1996. Orchideenkartierung Mallorca II. *Ber. Arbeitskreis. Heimische Orchid.* 13(1): 31-40.

Kranjčev, R. 2001. Orchids on the island of Vis (eastern Adriatic – Croatia). *Acta Bot. Croat.* 60: 69-74.

Kretzschmar, G. & Kretzschmar, H. 1996. Orchideen der Insel Naxos. *Ber. Arbeitskreis. Heimische Orchid.* 13(1): 4-30.

Kretzschmar, H. & Kretzschmar, G. 2003. Zur Variationsbreite von *Ophrys ferrum-equinum*. *Ber. Arbeitskreis. Heimische Orchid.* 20(1): 75-79.

Kretzschmar, H., Kretzschmar, G. & Eccarius, W. 2001. *Orchideen auf Rhodos. Ein Feldführer durch die Orchideenflora der "Insel des Lichts"*. Selbstverlag H. Kretzschmar, Bad Hersfeld.

Kretzschmar, H., Kretzschmar, G. & Eccarius, W. 2002. *Orchideen auf Kreta, Kasos und Karpathos. Ein Feldführer durch die Orchideenflora der zentralen Inseln der Südägäis*. Selbstverlag H. Kretzschmar, Bad Hersfeld.

Kretzschmar, H., Wenker, D. & Willing, E. 1984. Orchideenkartierung der Insel Rhodos – aktuelle Übersicht. *Ber. Arbeitskreis. Heimische Orchid.* 1: 130-146.

Kreutz, C. A. J. 1989. Beitrag zur Erforschung und Kenntnis einiger Orchideenarten der Balearen-Insel Mallorca. *Ber. Arbeitskreis. Heimische Orchid.* 6(1): 115-128.

Kreutz, C. A. J. 1990. Beitrag zur Orchideenflora Kretas. *Mitt. Arbeitskreis Heimische Orchid. Baden-Württemberg* 22: 358-384.

Kreutz, C. A. J. 1997. *Apochromen Ophrys-varianten van Midden-Europa*. Meijs Publishers, Limbricht.

Kreutz, C. A. J. 2001. *Ophrys helios*, eine neue Art von Karpathos (Ostägäis). *J. Eur. Orchid.* 33: 871-880.

Kreutz, C. A. J. 2002. *Die Orchideen von Rhodos und Karpathos. Beschreibung, Lebensweise, Verbreitung, Gefährdung, Schutz und Ikonographie*. Seckel & Kreutz Publishers, Raalte & Landgraaf.

Kullenberg, B. 1956. On the scents and colours of *Ophrys* flowers and their specific pollinators among the aculeate Hymenoptera. *Svensk Bot. Tidskr.* 50: 25-46.

Kullenberg, B. 1961. Studies in *Ophrys* pollination. *Zool. Bidrag Uppsala* 34: 1-340.

Kullenberg, B. 1973. New observations on the pollination of *Ophrys* L. (Orchidaceae). *Zoon*, suppl. 1: 9-13 & 2 pls.

Kullenberg, B. 1973a. Field experiments with chemical sexual attractants on aculeate Hymenoptera males. II. *Zoon*, suppl. 1: 31-42 & 2 pls.

Kullenberg, B. & Bergström, G. 1976. The pollination of *Ophrys* orchids. *Bot. Notiser* 129: 11-19.

Kullenberg, B., Borg-Karlson, A. K. & Kullenberg, A.-L. 1984. Field studies on the behaviour of the *Eucera nigrilabris* male in the odour flow from flower labellum extract of *Ophrys tenthredinifera*. *Nova Acta Reg. Soc. Sci. Ups., Ser.* V-C 3: 79-110.

Kullenberg, B., Büel, H. & Tkalců, B. 1984. Übersicht von Beobachtungen über Besuche von *Eucera*- und *Tetralonia*-Männchen auf *Ophrys*-Blüten (Orchidaceae). *Nova Acta Reg. Soc. Sci. Ups., Ser.* V-C 3: 27-40.

Künkele, S. & Lorenz, R. 1995. Zum Stand der Orchideenkartierung in Sizilien. Ein Beitrag zum OPTIMA-Projekt "Kartierung der mediterranen Orchideen". In: Senghas, K. & Lünsmann, U. (eds), 10. Wuppertaler Orchideen-Tagung. *Jahresber. Naturwiss. Vereins Wuppertal* 48: 21-115 (1-230 & 3 pls).

*Landwehr, J. 1977. *Wilde orchideeën van Europa* 1-2. Vereniging tot Behoud van Natuurmonumenten in Nederland, The Hague.

Lavarack, P. S. & Dixon, K. W. 2003. The role of legislation in orchid conservation. In: Dixon, K. W., Kell, S. P., Barrett, R. L. & Cribb, P. J. (eds), *Orchid conservation*. Natural History Publications (Borneo), Kota Kinabalu, pp. 289-298.

Levin, D. A. 1993. Local speciation in plants: the rule not the exception. *Syst. Bot.* 18: 197-208.

Levin, D. A. 2000. *The origin, expansion, and demise of plant species*. Oxford University Press, New York & Oxford.

Levin, D. A. 2001. The recurrent origin of plant races and species. *Syst. Bot.* 26: 197-204.

Lorella, B., Mahë, G. & Seité, F. 2002. Pollinisateurs d'*Ophrys* en Bretagne. *Orchidophile* 151: 91-96.

Lorenz, R. & Gembardt, C. 1987. Die Orchideenflora des Gargano (Italien). *Mitt. Arbeitskreis Heimische Orchid. Baden-Württemberg* 19: 385-756.

Lorenz, R. & Lorenz, K. 2002. Zur Orchideenflora zirkumsizilianischer Inseln. Ein Beitrag zum OPTIMA-Projekt "Kartierung der mediterranen Orchideen". *Jahresber. Naturwiss. Vereins Wuppertal* 55: 100-162.

Lowe, M. R. 1998. The orchids of the province of Málaga, Spain. (A contribution to the OPTIMA project "Mapping of Mediterranean Orchids"). *J. Eur. Orchid.* 30: 501-570.

Lucke, E. 1971. Zur Samenkeimung mediterraner *Ophrys*. *Orchidee (Hamburg)* 22: 62-65.

Malmgren, S. 2002. Growing *Ophrys* downstairs. *Hardy Orchid Soc. Newslett.* 24: 15–18.

Malmgren, S. 2004. On the origin of *Ophrys* species. *J. Hardy Orchid Soc.* 1: 74–81.

Malmgren, S. 2006. Asymbiotic propagation of European orchids. *J. Eur. Orchid.* 38: 355-362.

Manuel, R. L. 1989. Wild orchids of Rhodes. Part 1. *Orchid Rev.* 97: 4-10.

Manuel, R. L. 1989a. Wild orchids of Rhodes. Part 2. *Orchid Rev.* 97: 56-62.

Manuel, R. L. 1996–1997. Flasking forum. *Hardy Orchid Soc. Newslett.* 2: 4–9; 3: 10–13; 4: 6–10; 5: 19–23; 6: 8.

Manuel, R. L. 2005. Orchids outside. *J. Hardy Orchid Soc.* 2: 59–62.

Mitchell, R. B. 1989. Growing hardy orchids from seeds at Kew. *Plantsman* 2: 152–169.

Molero Mesa, J., Pérez Raya, F. A. & Martinez Parras, J. M. 1981. Relacíon de las Orchidaceae de la provincia de Granada. *Anales Jard. Bot. Madrid* 37: 645-659.

Möller, O. 1989. Der Grund für das Ausbleiben der Blüte bei *Ophrys apifera*: das Anfangsstadium des Knollenwuchses. *Orchidee (Hamburg)* 40: 61-64.

Möller, O. 1996. Das Wachstum der Ophrydeen und des *Cypripedium calceolus*. *Ber. Arbeitskreis. Heimische Orchid.* 13(2): 57-66.

Möller, O. 2000. Fotografische Dokumentation von Keimung und Aufwuchs der Erdorchideen in der Natur und die Darstellung dieses Geschehens in früherer Literatur. *Orchidee (Hamburg)* 51: 562-571.

Molnár, V. A. & Gulyás, G. 2005. Zur Kenntnis der *Ophrys holubyana* Andrasovszky 1917. *J. Eur. Orchid.* 37: 625-638.

*Mossberg, B. & Nilsson, S. 1987. *Orkidéer. Europas vildväxande arter.* Wahlström & Widstrand, Stockholm.

Mrkvicka, A. C. 1994. Anatomie und Morphologie der Samen heimischer Orchideenarten. *J. Eur. Orchid.* 26: 168-314.

Muir, H. J. 1989. Germination and mycorrhizal fungus compatibility in European orchids. In: Pritchard, H. W. (ed.), *Modern methods in orchid conservation: the role of physiology, ecology and management.* Cambridge University Press, Cambridge, pp. 39-56.

Nazarov, V. V. & Gerlach, G. 1997. The potential seed productivity of orchid flowers and peculiarities of their pollination systems. *Lindleyana* 12: 188-204.

Neiland, M. R. M. & Wilcock, C. C. 1995. Maximisation of reproductive success by European Orchidaceae under conditions of infrequent pollination. *Protoplasma* 187: 39-48.

*Nelson, E. 1962. *Gestaltwandel und Artbildung erörtert am Beispiel der Orchidaceen Europas und der Mittelmeerländer insbesondere der Gattung Ophrys – mit einer Monographie und Ikonographie der Gattung Ophrys.* Verlag E. Nelson, Chernex-Montreux.

Neto, N. B. M. & Custódio, C. C. 2005. Orchid conservation through seed banking: ins and outs. *Selbyana* 26: 229-235.

Paulus, H. F. 1988. Beobachtungen und Experimente zur Pseudokopulation auf *Ophrys*-Arten (Orchidaceae) Kretas (II) mit einer Beschreibung von *Ophrys sitiaca* H. F. Paulus & C. + A. Alibertis nov. spec. aus dem *Ophrys fusca-omegaifera*-Formenkreis. *Mitt. Arbeitskreis Heimische Orchid. Baden-Württemberg* 20: 817-882.

Paulus, H. F. 1994. Untersuchungen am *Ophrys cretica*-Komplex mit Beschreibung von *Ophrys ariadnae* H. F. Paulus spec. nov. (Orchidaceae). *J. Eur. Orchid.* 26: 629-643.

Paulus, H. F. 1996. Zur Bestäubungsbiologie und Artberechtigung von *Ophrys tetraloniae* Teschner 1987 und *Ophrys elatior* Gumprecht ex H. F. Paulus spec. nov. (Orchidaceae). *Ber. Arbeitskreis. Heimische Orchid.* 13(2): 4-13.

Paulus, H. F. 1998. Der *Ophrys fusca* s.str.-Komplex auf Kreta und anderer Ägäisinseln mit Beschreibungen von O. *blitopertha*, O. *creberrima*, O. *cinereophila*, O. *cressa*, O. *thriptiensis* und O. *creticola* spp. nov. (Orchidaceae). *J. Eur. Orchid.* 30: 157-201.

Paulus, H. F. 2000. Zur Bestäubungsbiologie einiger *Ophrys*-Arten Istriens (Kroatien) mit einer Beschreibung von *Ophrys serotina* Rolli ex Paulus spec. nov. aus der *Ophrys holoserica*-Artengruppe (Orchidaceae und Insecta, Apoidea). *Ber. Arbeitskreis. Heimische Orchid.* 17(2): 4-33.

Paulus, H. F. 2001. Material zu einer Revision des *Ophrys fusca* s.str. Artenkreises I. – *Ophrys nigroaena-fusca*, O. *colletes-fusca*, O. *funerea*, O. *forestieri* oder was ist die typische *Ophrys fusca* Link 1799 (Orchidaceae)? *J. Eur. Orchid.* 33: 121-177.

Paulus, H. F. 2001a. *Andrena paucisquama*-Männchen als Bestäuber der Sexualtäuschblume *Ophrys aesculapii* Renz 1928 (Orchidaceae; Insecta, Apoidea: Andrenidae). *Ber. Arbeitskreis. Heimische Orchid.* 18(1): 4-10.

Paulus, H. F. 2001b. Daten zur Bestäubungsbiologie und Systematik der Gattung *Ophrys* in Rhodos (Griechenland) mit Beschreibung von *Ophrys parvula*, *Ophrys persephonae*, *Ophrys lindia*, *Ophrys eptapigiensis* spec. nov. aus der *Ophrys fusca* s.str. Gruppe und

Ophrys cornutula spec. nov. aus der *Ophrys oestrifera*-Gruppe (Orchidaceae und Insecta, Apoidea). *Ber. Arbeitskreis. Heimische Orchid.* 18(1): 38-86.

Paulus, H. F. 2002. Daten zur Bestäubungsbiologie und Systematik der Gattung *Ophrys* in Rhodos (Griechenland) II. Über *Ophrys holoserica* s.lat.: *Ophrys episcopalis, Op. maxima* und *Ophrys halia* spec. nov. (Orchidaceae und Insecta, Apoidea). *Ber. Arbeitskreis. Heimische Orchid.* 18(2): 46-63.

Paulus, H. F. 2006. Deceived males – pollination biology of the Mediterranean orchid genus *Ophrys* (Orchidaceae). *J. Eur. Orchid.* 38: 303-353.

Paulus, H. F., Alibertis, C. & Alibertis, A. 1990. *Ophrys mesaritica* H. F. Paulus und C. & A. Alibertis spec. nov. aus Kreta, eine neue Art aus dem *Ophrys fusca-iricolor*-Artenkreis. *Mitt. Arbeitskreis Heimische Orchid. Baden-Württemberg* 22: 772-787.

Paulus, H. F. & Gack, C. 1980. Beobachtungen und Untersuchungen zur Bestäubungsbiologie südspanischer *Ophrys*-Arten. In: Senghas, K. & Sundermann, H. (eds), Probleme der Evolution bei europäischen und mediterranen Orchideen. *Orchidee (Hamburg)*, special issue November 1980: 55-68 (1-178 & 2 pls).

Paulus, H. F. & Gack, C. 1981. Neue Beobachtungen zur Bestäubung von *Ophrys* (Orchidaceae) in Südspanien, mit besonderer Berücksichtigung des Formenkreises *Ophrys fusca* agg. *Pl. Syst. Evol.* 137: 241-258.

Paulus, H. F. & Gack, C. 1983. Untersuchungen zur Bestäubung des *Ophrys fusca*-Formenkreises in Südspanien. Ein Beitrag zum Biospecies-Konzept der Gattung *Ophrys*. In: Senghas, K. & Sundermann, H. (eds), Probleme der Taxonomie, Verbreitung und Vermehrung europäischer und mediterraner Orchideen. *Orchidee (Hamburg)*, special issue March 1983: 65-72 (1-120).

Paulus, H. F. & Gack, C. 1986. Neue Befunde zur Pseudokopulation und Bestäuberspezifität in der Orchideengattung *Ophrys* – Untersuchungen in Kreta, Süditalien und Israel. In: Senghas, K. & Sundermann, H. (eds), Probleme der Taxonomie, Verbreitung und Vermehrung europäischer und mediterraner Orchideen, II. *Orchidee (Hamburg)*, special issue May 1986: 48-86 (1-160).

Paulus, H. F. & Gack, C. 1990. Pollinators as prepollinating isolation factors: evolution and speciation in *Ophrys* (Orchidaceae). *Israel J. Bot.* 39: 43-79.

Paulus, H. F. & Gack, C. 1990a. Pollination of *Ophrys* (Orchidaceae) in Cyprus. *Pl. Syst. Evol.* 169: 177-207.

Paulus, H. F. & Gack, C. 1990b. Untersuchungen zur Pseudokopulation und Bestäuberspezifität in der Gattung *Ophrys* im östlichen Mittelmeergebiet (Orchidaceae, Hymenoptera, Apoidea). In: Senghas, K., Sundermann, H. & Kolbe, W. (eds), Probleme bei europäischen und mediterranen Orchideen. *Jahresber. Naturwiss. Vereins Wuppertal* 43: 80-118 (1-176 & 2 pls).

Paulus, H. F. & Gack, C. 1990c. Zur Pseudokopulation und Bestäuberspezifität der Gattung *Ophrys* in Sizilien und Süditalien. In: Senghas, K., Sundermann, H. & Kolbe, W. (eds), Probleme bei europäischen und mediterranen Orchideen. *Jahresber. Naturwiss. Vereins Wuppertal* 43: 119-141 (1-176 & 2 pls).

Paulus, H. F. & Gack, C. 1992. Die Gattung *Ophrys* (Orchidaceae) auf der Kykladeninsel Naxos: Daten zur Bestäubungsbiologie und zur Floristik. *Mitt. Arbeitskreis Heimische Orchid. Baden-Württemberg* 24: 403-449.

Paulus, H. F. & Gack, C. 1995. Zur Pseudokopulation und Bestäubung in der Gattung *Ophrys* (Orchidaceae) Sardiniens und Korsikas. In: Senghas, K. & Lünsmann, U. (eds), 10. Wuppertaler Orchideen-Tagung. *Jahresber. Naturwiss. Vereins Wuppertal* 48: 188-227 (1-230 & 3 pls).

Paulus, H. F. & Gack, C. 1999. Bestäubungsbiologische Untersuchungen an der Gattung *Ophrys* in der Provence (SO-Frankreich), Ligurien und Toscana (NW-Italien) (Orchidaceae und Insecta, Apoidea). *J. Eur. Orchid*. 31: 347-422.

Paulus, H. F., Gack, C. & Maddocks, R. 1983. Beobachtungen und Experimente zum Pseudokopulationsverhalten an *Ophrys* – das Lernverhalten von *Eucera barbiventris* an *Ophrys scolopax* in Südspanien. In: Senghas, K. & Sundermann, H. (eds), Probleme der Taxonomie, Verbreitung und Vermehrung europäischer und mediterraner Orchideen. *Orchidee (Hamburg)*, special issue March 1983: 73-79 (1-120).

Pedersen, H. Æ. 1998. Species concept and guidelines for infraspecific taxonomic ranking in *Dactylorhiza* (Orchidaceae). *Nordic J. Bot*. 18: 289-310.

Pedersen, H. Æ. & Faurholdt, N. 1997. A critical approach to *Ophrys calypsus* (Orchidaceae) and to the records of *O. holoserica* subsp. *apulica* from Greece. *Flora Medit*. 7: 153-162.

Pedersen, H. Æ. & Faurholdt, N. 1997a. Beiträge zur Orchideenflora der ostmediterranen Inseln Rhodos und Zypern. *Orchidee (Hamburg)* 48: 232-236.

Pedersen, H. Æ. & Faurholdt, N. 2002. *Ophrys* – versuchsweise Definitionen der Kategorien Art, Unterart und Varietät in der Gattung und einige daraus resultierende taxonomische Änderungen. *Orchidee (Hamburg)* 53: 341-346.

Peitz, E. 1967. *Gestalt- und Farb-Abwandlungen an Orchideen, insbesondere bei Ophrys fuciflora*. Heimvolkshochschule Schloß Dhaun, Dhaun.

Peter, R. 1989. Ergänzungen zur Orchideenflora von Rhodos. *Mitt. Arbeitskreis Heimische Orchid*. 21: 279-350.

Petersen, B. V. & Høyer-Nielsen, J. 2005. *Ophrys apifera* Huds. – nu i Danmark. *Urt* 29: 16-19.

Pridgeon, A. M., Bateman, R. M., Cox, A. V., Hapeman, J. R. & Chase, M. W. 1997. Phylogenetics of subtribe Orchidinae (Orchidoideae, Orchidaceae) based on nuclear ITS sequences. 1. Intergeneric relationships and polyphyly of *Orchis* sensu lato. *Lindleyana* 12: 89-109.

Pridgeon, A. M., Cribb, P. J., Chase, M. W. & Rasmussen, F. N. (eds) 1999. *Genera Orchidacearum 1. General introduction, Apostasioideae, Cypripedioideae*. Oxford University Press, Oxford.

Pridgeon, A. M., Cribb, P. J., Chase, M. W. & Rasmussen, F. N. (eds) 2001. *Genera Orchidacearum 2. Orchidoideae (part 1)*. Oxford University Press, Oxford.

Priesner, E. 1973. Reaktionen von Riechrezeptoren männlicher Solitärbienen (Hymenoptera, Apoidea) auf Inhaltsstoffe von *Ophrys*-Blüten. *Zoon*, suppl. 1: 43-54 & 1 pl.

Ramsay, M. M. & Dixon, K. W. 2003. Propagation science, recovery and translocation of terrestrial orchids. In: Dixon, K. W., Kell, S. P., Barrett, R. L. & Cribb, P. J. (eds), *Orchid conservation*. Natural History Publications (Borneo), Kota Kinabalu, pp. 259-288.

Ramsay, M. M., Stewart, J. & Prendergast, G. 1994. Conserving endangered British orchids. In: Pridgeon, A. M. (ed.), *Proceedings of the 14th World Orchid Conference*. HMSO, London, pp. 176-179.

Rasmussen, H. 1995. *Terrestrial orchids from seed to mycotrophic plant*. Cambridge University Press, Cambridge.

Reinecke, F. 1995. Über die Ausbringung von Orchideen als Mittel zum Arterhalt. In: Senghas, K. & Lünsmann, U. (eds), 10. Wuppertaler Orchideen-Tagung. *Jahresber*.

Naturwiss. Vereins Wuppertal 48: 116-131 (1-230 & 3 pls).

Reinhard, H. R. 1987. Untersuchungen an *Ophrys holoserica* (Burm. fil.) W. Greuter subsp. *elatior* (Gumprecht) Gumprecht (Orchidaceae). *Mitt. Arbeitskreis Heimische Orchid. Baden-Württemberg* 19: 769-800.

Renz, J. 1980. Entwicklungstendenzen bei den Orchidoideae – einige Betrachtungen. In: Senghas, K. & Sundermann, H. (eds), Probleme der Evolution bei europäischen und mediterranen Orchideen. *Orchidee (Hamburg)*, special issue November 1980: 35-43 (1-178 & 2 pls).

Richard, D. & Evans, D. 2006. The need for plant taxonomy in setting priorities for designated areas and conservation management plans: a European perspective. In: Leadley, E. & Jury, S. (eds.), *Taxonomy and plant conservation: the cornerstone of the conservation and the sustainable use of plants*. Cambridge University Press, Cambridge, pp. 162-176.

Riechelmann, A. 2006. Orchideenexkursion Andalusien (Provinz Malaga) vom 20. März bis 03. April 2005. *Ber. Arbeitskreis. Heimische Orchid.* 22(2): 53-63.

Rieseberg, L. H., Church, S. A. & Morjan, C. L. 2004. Integration of populations and differentiation of species. *New Phytol.* 161: 59-69.

Roberts, D. L. 2003. Pollination biology: the role of sexual reproduction in orchid conservation. In: Dixon, K. W., Kell, S. P., Barrett, R. L. & Cribb, P. J. (eds), *Orchid conservation*. Natural History Publications (Borneo), Kota Kinabalu, pp. 113-136.

Ronse, A. 1989. In vitro propagation of orchids and nature conservation: possibilities and limitations. *Mém. Soc. Roy. Bot. Belg.* 11: 107-114.

★Rossi, W. 2002. *Orchidee d'Italia*. Ministero dell'Ambiente & Instituto Nazionale per la Fauna Selvatica.

Rossini, A. & Quitadamo, G. 1996. *Le Orchidee del Gargano*. Leone Editrice, Foggia.

Röttger, B. 1990. Beiträge zur Kartierung der Orchideenflora von Rhodos. *Mitt. Arbeitskreis Heimische Orchid. Baden-Württemberg* 22: 387-404.

Rudall, P. J. & Bateman, R. M. 2002. Roles of synorganization, zygomorphy and heterotopy in floral evolution: the gynostemium and labellum of orchids and other lilioid monocots. *Biol. Rev.* 77: 403-441.

Salanon, R. & Kulesza, V. 1998. *Mémento de la flore protégée des Alpes-Maritimes*. Office National des Forêts, Paris.

Saliaris, P. A. 2002. *Oi orchidees thes Chioy*. Ekdose Dimos Kardamilon, Chios.

Salkowski, H.-E. 2000. Über die Bestäubung von *Ophrys melitensis* (Salkowski) J. & P. Devillers-Terschuren auf der Insel Malta. *J. Eur. Orchid.* 32: 631-641.

Sanger, N. P. & Waite, S. 1998. The phenology of *Ophrys sphegodes* (the early spider orchid): what annual censuses can miss. In: Waite, S. (ed.), Orchid population biology: conservation and challenges. *Bot. J. Linn. Soc.* 126: 75-81 (1-190).

Schick, B. & Seack, K.-H. 1988. Rostelldifferenzierung und Pollarienbildung europäischer Orchideen V. Kinematographische Dokumentation der manuellen Bestäubung einer *Ophrys*-Blüte (*Ophrys insectifera*). *Orchidee (Hamburg)* 39: 27-32.

Schiestl, F. P. & Ayasse, M. 2001. Post-pollination emission of a repellent compound in a sexually deceptive orchid: a new mechanism for maximizing reproductive success? *Oecologia* 126: 531-534.

Schiestl, F. P. & Ayasse, M. 2002. Do changes in floral odor cause speciation in sexually deceptive orchids? *Pl. Syst. Evol.* 234: 111-119.

Schiestl, F. P., Ayasse, M., Paulus, H. F., Erdmann, D. & Francke, W. 1997. Variation of floral scent emission and postpollination changes in individual flowers of *Ophrys sphegodes* subsp. *sphegodes*. *J. Chem. Ecol.* 23: 2881-2895.

Schiestl, F. P., Ayasse, M., Paulus, H. F., Löfstedt, C., Hansson, B. S., Ibarra, F. & Francke, W. 1999. Orchid pollination by sexual swindle. *Nature* 399: 421-422.

Schiestl, F. P., Ayasse, M., Paulus, H. F., Löfstedt, C., Hansson, B. S., Ibarra, F. & Francke, W. 2000. Sex pheromone mimicry in the early spider orchid (*Ophrys sphegodes*): patterns of hydrocarbons as the key mechanism for pollination by sexual deception. *J. Comp. Physiol.* A 186: 567-574.

Schmid, R. & Schmid, M. J. 1977. Fossil history of the Orchidaceae. In: Arditti, J. (ed.), *Orchid biology. Reviews and perspectives, I.* Comstock Publishing Associates, Ithaca & London, pp. 25-45.

Schot, M. H. 1997. Orchideeënvariaties op Rhodos. *Eurorchis* 9: 21-32.

Scrugli, A. 1990. *Orchidee spontanee della Sardegna.* Edizioni Della Torre, Cagliari.

Scrugli, A. & Cogoni, A. 1998. Sardinia's orchids: taxonomic and phytogeographic considerations. *Caesiana* 11: 1-26.

Seaton, P. & Ramsay, M. 2005. *Growing Orchids from Seed.* Royal Botanic Gardens, Kew.

Selisky, H. 2002. *Ophrys aegaea* auf den Kykladeninseln Amorgos und Iraklia. *J. Eur. Orchid.* 34: 739-742.

Servettaz, O., Bini Maleci, L. & Grünanger, P. 1994. Labellum micromorphology in the *Ophrys bertolonii* agg. and some related taxa (Orchidaceae). *Pl. Syst. Evol.* 189: 123-131.

Singer, R. B. 2002. The pollination mechanism in *Trigonidium obtusum* Lindl. (Orchidaceae: Maxillariinae): sexual mimicry and trap-flowers. *Annals Bot.* 89: 157-163.

Singer, R. B., Flach, A., Koehler, S., Marsaioli, A. J. & Amaral, M. D. C. E. 2004. Sexual mimicry in *Mormolyca ringens* (Lindl.) Schltr. (Orchidaceae: Maxillariinae). *Annals Bot.* 93: 755-762.

Soliva, M., Kocyan, A. & Widmer, A. 2001. Molecular phylogenetics of the sexually deceptive orchid genus *Ophrys* (Orchidaceae) based on nuclear and chloroplast DNA sequences. *Molec. Phylogenet. Evol.* 20: 78-88.

*Souche, R. 2004. *Les orchidées sauvages de France.* Les Créations du Pélican, Paris.

Stahl, H. 1989. Zur Entwicklung einer Population von *Ophrys apifera* im Gebiet von Stuttgart. *Mitt. Arbeitskreis Heimische Orchid. Baden-Württemberg* 21: 1015-1039.

Stahl, H. 1993. Zur Entwicklung einer Population von *Ophrys apifera* im Gebiet von Stuttgart (Teil 2). *Mitt. Arbeitskreis Heimische Orchid. Baden-Württemberg* 25: 368-384.

van Steenis, C. G. G. J. 1957. Specific and infraspecific delimitation. In: van Steenis, C. G. G. J. (ed.) 1955-1958, *Flora Malesiana ser. I, Spermatophyta 5.* Noordhoff-Kolff N.V., Jakarta, pp. CLXVII-CCXXXIV.

Stern, W. L. 1997. Vegetative anatomy of subtribe Orchidinae (Orchidaceae). *Bot. J. Linn. Soc.* 124: 121-136.

Stewart, J. 1987. Orchid conservation: survival and maintenance of genetic diversity of all orchids throughout the world. *Amer. Orchid Soc. Bull.* 56: 822-827.

Stewart, J. 1992. Research and conservation. In: Stewart, J. (ed.), *Orchids at Kew.* The Royal Botanic Gardens, Kew & HMSO, London, pp. 103-138.

Stone, D. A. & Taylor, P. E. 1999. *Ophrys fuciflora* (Crantz) Moench & Reichenb. (Orchidaceae). In: Wigginton, M. J. (ed.), *British red data books 1. Vascular plants.* 3rd ed. Joint Nature Conservation Committee, Peterborough, p. 258.

Stoutamire, W. P. 1974. Australian terrestrial orchids, thynnid wasps, and pseudocopulation. *Amer. Orchid Soc. Bull.* 43: 13-18.

Stoutamire, W. P. 1975. Pseudocopulation in Australian terrestrial orchids. *Amer. Orchid Soc. Bull.* 44: 226-233.

Summerhayes, V. S. 1951. *Wild orchids of Britain with a key to the species.* Collins, London.

Sundermann, H. 1961. Standorte europäischer Orchideen. I. Gliederung in Standorttypen. *Orchidee (Hamburg)* 12: 131-135, 137.

Sundermann, H. 1962. Standorte europäischer Orchideen. II. Die Halbtrockenrasen (Mesobrometen) des Kaiserstuhlgebietes. *Orchidee (Hamburg)* 13: 5-9.

Sundermann, H. 1962a. Standorte europäischer Orchideen. IV. Mittelmeergebiet. *Orchidee (Hamburg)* 13: 125-132.

Sundermann, H. 1962b. Standorte europäischer Orchideen. V. Orchideenwälder. *Orchidee (Hamburg)* 13: 205-211.

Sundermann, H. 1964. Zum Problem der Artabgrenzung innerhalb der Gattung *Ophrys.* In: Sundermann, H. & Haber, W. (eds), Probleme der Orchideengattung *Ophrys. Orchidee (Hamburg),* special issue March 1964: 9-17 (1-72 & 6 pls).

Sundermann, H. 1972. Artenproduktion und Konsumbedürfnis. Kritische Bemerkungen zur Benennung und Beschreibung von Splittersippen. *Orchidee (Hamburg)* 23: 166-168.

Sundermann, H. 1975. Zum Problem der Definition taxonomischer Kategorien (Spezies, Subspezies, Praespezies, Varietät) – dargestellt am Beispiel der Sippenkomplexes *Ophrys fuciflora* (Crantz) Moench – *Ophrys scolopax* Cav. *Taxon* 24: 615-627.

Sundermann, H. 1976. Kritische Bemerkungen zur Bedeutung der Hybridisierung für die Artbildung – ein Diskussionsbeitrag. In: Senghas, K. (ed.), *Tagungsbericht der 8. Welt-Orchideen-Konferenz Palmengarten Frankfurt 10-17 April 1975.* Deutsche Orchideen Gesellschaft e.V., Frankfurt am Main, pp. 123-125.

Sundermann, H. 1977. The genus *Ophrys* – an example of the importance of isolation for speciation. *Amer. Orchid Soc. Bull.* 46: 825-831.

*Sundermann, H. 1980. *Europäische und mediterrane Orchideen. Eine Bestimmungsflora mit Berücksichtigung der Ökologie.* 3rd ed. Brücke-Verlag Kurt Schmersow, Hildesheim.

Sundermann, H. 1987. Kritische Bemerkungen zum Konzept von Baumann und Künkele. *Mitt. Arbeitskreis Heimische Orchid. Baden-Württemberg* 19: 97-107.

Szlachetko, D. L. & Rutkowski, P. 2000. Gynostemia Orchidalium I. Apostasiaceae, Cypripediaceae, Orchidaceae (Thelymitroideae, Orchidoideae, Tropidioideae, Spiranthoideae, Neottioideae, Vanilloideae). *Acta Bot. Fenn.* 169: 1-380.

Taylor, M. 2005. *Illustrated checklist. Orchids of Chios, Inouses & Psara.* "Pelineo" Centre of Chian Studies, Chios.

Thompson, P. A. 1977. *Orchids from Seed.* Royal Botanic Gardens, Kew.

Tyteca, D. 1997. The orchid flora of Portugal. *J. Eur. Orchid.* 29: 185-581.

Tyteca, D. 2000. The orchid flora of Portugal – addendum n. 3 – remarks on *Spiranthes spiralis* (L.) Chevall. and three new taxa to the Portuguese flora. *J. Eur. Orchid.* 32: 291-347.

Tyteca, D. & Tyteca, B. 1986. Orchidées du Portugal – 11. Esquisse systématique, chorologique et cartographique. *Naturalistes Belges* 67: 163-192.

Vöth, W. 1984. Bestäubungsbiologische Beobachtungen an griechischen *Ophrys*-Arten. *Mitt. Arbeitskreis Heimische Orchid. Baden-Württemberg* 16: 1-20.

Vöth, W. 1985. Ermittlung der Bestäuber von *Ophrys fusca* Link subsp. *funerea* (Viv.) G. Camus, Bergon & A. Camus und von *Ophrys lutea* Cav. subsp. *melena* Renz. *Mitt.*

Arbeitskreis Heimische Orchid. Baden-Württemberg 17: 417-445.

Vöth, W. 1986. Zum Nachweis des Bestäubers *Melecta albifrons albovaria* Erichs. von *Ophrys cretica* (Vierh.) Nelson auf der griechischen Insel Aejina. *Mitt. Arbeitskreis Heimische Orchid. Baden-Württemberg* 18: 243-253.

Vöth, W. 1987. Neue bestäubungsbiologische Beobachtungen an griechischen *Ophrys*-Arten. *Mitt. Arbeitskreis Heimische Orchid. Baden-Württemberg* 19: 112-118 & 2 pls.

Vöth, W. & Ehrendorfer, F. 1976. Biometrischer Untersuchungen an Populationen von *Ophrys cornuta*, *O. holoserica* und ihren Hybriden (Orchidaceae). *Pl. Syst. Evol.* 124: 279-290.

Vöth, W. & Löschl, E. 1978. Zur Verbreitung der Orchideen an der östlichen Adria. *Linzer Biol. Beitr.* 10: 369-430.

Waite, S. 1989. Predicting population trends in *Ophrys sphegodes* Mill. In: Pritchard, H. W. (ed.), *Modern methods in orchid conservation: the role of physiology, ecology and management*. Cambridge University Press, Cambridge, pp. 117-126.

Waite, S. & Hutchings, M. J. 1991. The effects of different management regimes on the population dynamics of *Ophrys sphegodes*: analysis and description using matrix models. In: Wells, T. C. E. & Willems, J. H. (eds), *Population ecology of terrestrial orchids*. SPB Academic Publishing bv, The Hague, pp. 161-175.

Walravens, É. 1995. Un pollinisateur pour *Ophrys aurelia* P. Delforge, J. & P. Devillers-Terschuren 1989. *Naturalistes Belges* 76: 98-102.

Warncke, K. & Kullenberg, B. 1984. Übersicht von Beobachtungen über Besuche von *Andrena-* und *Colletes cunicularius*-Männchen auf *Ophrys*-Blüten (Orchidaceae). *Nova Acta Reg. Soc. Sci. Ups., Ser.* V-C 3: 41-55.

Wellinghausen, N. & Koch, H. 1989. Orchideensuche auf Kreta. *Ber. Arbeitskreis. Heimische Orchid.* 6: 85-100.

Wells, T. C. E. & Cox, R. 1989. Predicting the probability of the bee orchid (*Ophrys apifera*) flowering or remaining vegetative from the size and number of leaves. In: Pritchard, H. W. (ed.), *Modern methods in orchid conservation: the role of physiology, ecology and management*. Cambridge University Press, Cambridge, pp. 127-139.

Wells, T. C. E. & Cox, R. 1991. Demographic and biological studies on *Ophrys apifera*: some results from a 10 year study. In: Wells, T. C. E. & Willems, J. H. (eds), *Population ecology of terrestrial orchids*. SPB Academic Publishing bv, The Hague, pp. 47-61.

Wiefelspütz, W. 1964. Über die Selbstbefruchtung bei *Ophrys apifera*. In: Sundermann, H. & Haber, W. (eds), Probleme der Orchideengattung *Ophrys*. *Orchidee (Hamburg)*, special issue March 1964: 56-62 (1-72 & 6 pls).

Wirth, W. & Blatt, H. 1988. Kritische Anmerkungen zu "Die Gattung *Ophrys* L. eine taxonomische Übersicht". *Ber. Arbeitskreis. Heimische Orchid.* 5(1/2): 4-21.

Wolff, T. 1950. Pollination and fertilisation of the fly ophrys, *Ophrys insectifera* L. in Allindelille Fredskov, Denmark. *Oikos* 2: 20-59.

Wolff, T. 1951. Ecological investigations on the fly ophrys, *Ophrys insectifera* L. in Allindelille Fredskov, Denmark. *Oikos* 3: 71-97.

Wood, J. J. 1982. Eine neue Subspecies von *Ophrys holoserica* aus Sardinien und eine neue interspezifische Hybride von Parma, Italien. *Orchidee (Hamburg)* 33: 66-69.

Ziegenspeck, H. 1936. Orchidaceae. In: von Kirchner, O., Loew, E., Schröter, C. & Wangerin, W. (eds), *Lebensgeschichte der Blütenpflanzen Mitteleuropas. Spezielle Ökologie der Blütenpflanzen Deutschlands, Österreichs und der Schweiz. I, 4*. Verlagsbuchhandlung Eugen Ulmer, Stuttgart, pp. I-VII, 1-840.

Index to scientific names of orchids

Accepted names are in Roman type. Synonyms are in *italics*. Numbers in **bold** refer to main taxonomic accounts.

Index to scientific names of insects

Abbreviations of geographic names

Europe

Aeg The eastern Greek islands in the Aegean Sea, north to (and including) Lesbos

Alb Albania

Aus Austria/Liechtenstein

Bal The Balearic Islands

Bel Belgium

Bul Bulgaria

Can The Canary Islands

Cor Corsica

Cre Crete/Karpathos/Kasos/Gavdhos

Cze Czech Republic/Slovakia

Den Denmark

Eng England

Est Estonia

Fin Finland

Fra France (excluding "Cor")

Ger Germany

Gre Greece (excluding "Aeg", "Cre")

Hol Holland

Hun Hungary

Ire Ireland

Ita Italy (excluding "Sar", "Sic")

Lat Latvia

Lit Lithuania

Lux Luxemburg

Mal Malta

Nor Norway

Pol Poland

Por Portugal

Rum Rumania

Rus Russia

Sar Sardinia

Sco Scotland

Sic Sicily and nearby islets

Spa   Spain (excluding "Bal", "Can")/Andorra
Swe   Sweden
Swi   Switzerland
Tur   Turkey north of the Sea of Marmara/the Bosporus
Ukr   The Ukraine
Wal   Wales
Yug   Yugoslavia as delimited by December 1990

Outside Europe

Ana   Anatolia
Cyp   Cyprus
Isr   Israel
Mor   Morocco
Tun   Tunisia

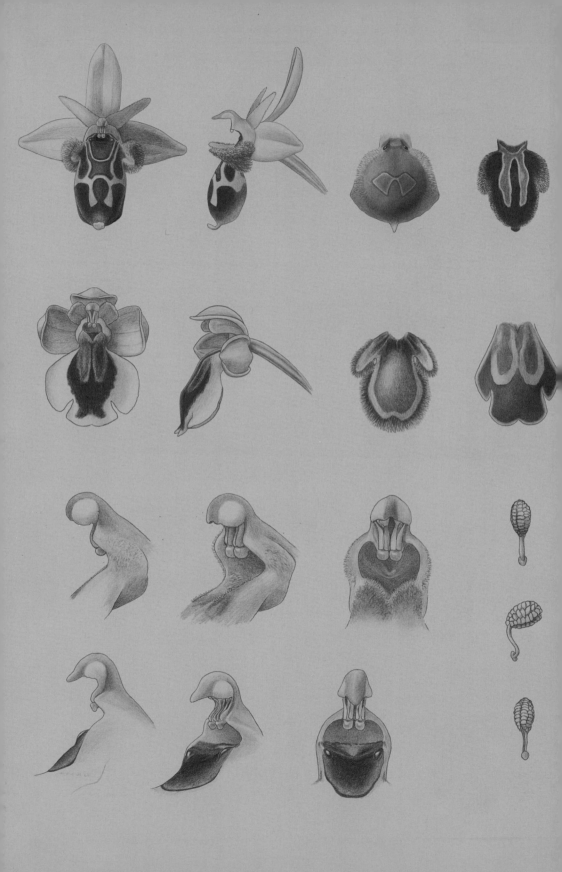